Enjoy Writing Your Science Thesis

or Dissertation!

Imperial College Press

Enjoy Writing Your Science Thesis

or Dissertation!

A step by step guide to planning and writing dissertations and theses for undergraduate and graduate science students

Daniel Holtom & Elizabeth Fisher

Imperial College Press

Published by

Imperial College Press
57 Shelton Street
Covent Garden
London WC2H 9HE

Distributed by

World Scientific Publishing Co. Pte. Ltd.
5 Toh Tuck Link, Singapore 596224
USA office: Suite 202, 1060 Main Street, River Edge, NJ 07661
UK office: 57 Shelton Street, Covent Garden, London WC2H 9HE

Library of Congress Cataloging-in-Publication Data
Holtom, Daniel.
    Enjoy writing your science thesis or dissertation : a step by step
guide to planning and writing dissertations and theses for
undergraduate and graduate science students / Daniel Holtom and
Elizabeth Fisher.
      p.   cm.
    Includes bibliographical references (p.   ) and index.
    ISBN 1-86094-090-0 -- ISBN 1-86094-207-5 (pbk)
    1. Technical writing.   2. Dissertations, Academic.   I. Fisher,
Elizabeth.   II. Title.
    T11.H582   1999
    808'.0666--dc21   99-11015

British Library Cataloguing-in-Publication Data
A catalogue record for this book is available from the British Library.

First published 1999
Reprinted 2003

Printed in Singapore.

Enjoy Writing Your Science Thesis Or Dissertation!

Almost all science students have to write a thesis or dissertation at some point during their careers. For undergraduate and masters students, assessment of the thesis is a component of the final mark; in the case of doctoral students the degree almost entirely depends on it. The skill of thesis writing lies in the clear organisation of data, ideas and resources. Unfortunately, many science students get very little formal training in the skill of communicating ideas clearly and efficiently, and as a result they lose marks because of poorly planned and presented work.

Enjoy Writing Your Science Thesis Or Dissertation! is the complete guide to good thesis or dissertation writing: giving practical advice and taking the student through the process of planning, writing, editing, presenting, and submitting a successful thesis. It gives the reader a clear understanding of both the theory and the practice of writing a scientific thesis or dissertation. It provides an organisational framework for the student and extensive backup information, including a guide to the pitfalls of scientific and general English for both native and non-native English speakers. For quick reference, reminders of key points are given at the end of most chapters.

Enjoy Writing Your Science Thesis Or Dissertation! is a guide for use either from day one of the degree course—or as a life jacket for a floundering student with a looming deadline.

Enjoy Writing Your Science Thesis Or Dissertation! is written by a professional writer with experience of teaching English, and by an academic scientist with experience of examining and supervising undergraduate and graduate projects. This guide is aimed at science students at all levels, and also provides useful pointers for career scientists on scientific writing in general.

We wish to thank many people for helpful comments and advice on this manuscript, in particular Julian Taylor, Don Doering, Owen Garling, Ramsay Gohar, David Goodgame, Jonathan Halliwell, Simon Leather, Tonya Pillie, Marek Sergot, Johnny and Chrissie Pillie-Rabbs, Alex Souper, Nessan Bermingham, Richard Halls, Diana Hernandez, Jacqui Hoyle, Nico Katsanis, Colin Kerr, Anna Kessling, John King, Sharon Nicholson, Abi Witherden.

The views and opinions expressed in this book are those of the authors. We welcome any comments and stories from students and supervisors alike for future revisions of this guide. All the anecdotes in this book are true and involve real people all though we have not mentioned any names...

CONTENTS

Chapter 1 provides basic information and summarises how to organise, plan and write a thesis or dissertation. The remaining chapters go into much greater detail on each topic; because each chapter can be read independently of the main text you will find repetition between them. Use what is helpful to you.

We recommend that you read Chapter 1 as a whole, and then use the rest of the book as a reference section.

Chapter 8
Deciding On a Title and Planning and Writing the
Other Bits

Chapter 9
Proofreading, Printing, Binding and Submission

Chapter 10
And You Thought It Was All Over

Chapter 1

ENJOY WRITING YOUR SCIENCE THESIS OR DISSERTATION!

We are assuming you have more than enough reading to do and possibly very little time. This chapter is a short overall guide with information summarised under useful headings. It is meant to be read as a whole and will give you a basic understanding of the standard conventions of dissertation and thesis writing, along with suggestions as to the best way of approaching the task. It will help you streamline the process of producing a rational and readable dissertation, thesis, or yearly report for a BSc, BA, BEng, MSc, MPhil, MEng, PhD, DPhil—or any other degree.

The rest of the book provides more detailed advice about the different aspects of thesis and dissertation writing. Read the parts that are useful to you, take what is helpful and leave what is not.

We are also assuming that you will type your thesis on a computer, that you have some basic word processing skills, and that you have access to a library in which you can carry out literature searches (see *Chapter 12: Resources*).

What are the Rules?

There are very few rules for writing a dissertation or thesis. The rules that do exist mainly concern the formatting of your work (number of copies, layout, Title Page, Abstract or Summary, binding and so on) and these differ between departments and universities. Find out your local rules as soon as possible.

1

While there are few rules, there are many conventions—modes of writing that we all conform to, usually because they are efficient ways of conveying information. Each discipline has its own conventions for the structure of theses and dissertations. The best way of finding out about the conventions in your area of research, and your department is to ask your supervisor to recommend a good recent dissertation or thesis in your own field as a guide to form, content and style.

Structure

As with any other story, the structure of a scientific thesis or dissertation has three parts: the beginning, the middle, and the end. In its final form your dissertation or thesis will probably be laid out something like this:

The Beginning:
    Title Page
    Abstract
    Dedication
    Acknowledgements
    Table of Contents
    List of Figures
    List of Tables
    List of Appendices
    List of Abbreviations (also known as 'Nomenclature' in some disciplines)
    Introduction (including a literature review)
The Middle:
    Materials and Methods/Experimental Techniques
    Results
The End:
    Discussion
    References (also known as 'Bibliography' in some disciplines)
    Glossary
    Appendices
    Published Papers

You probably will not have all these sections in your thesis or dissertation, but this scheme provides a basic structure from which to plan your writing.

Thesis and Dissertation Templates

Some departments have 'thesis templates' available over the World Wide Web or on floppy disk. These give you the basic structure of the thesis (Title Page, Abstract, Introduction, Results pages, etc.) all you have to do is to fill in the blanks with your own text. Before you start planning your thesis or dissertation, find out if there is a template that you should be using (see *Chapter 12: Resources*).

The Importance of Planning

A dissertation or thesis is not simply a list of experiments with a vague outline of what they all mean. The text needs a clear structure that starts by introducing the reader to the topic, states the aim of the research, shows the results and then discusses their significance. You need a plan. Without one, it is easy to overlook important points or jump about randomly from idea to idea.

This guide is written on the assumption that you will first develop a complete plan of your thesis or dissertation. Only when this is in place and you can see the structure of your text and have defined your aim, should you start writing.

Familiarise Yourself With the Appropriate Format and Style

A useful first exercise is to have a careful look through a few recent theses in your field. Browse through them to see how the work has been divided into sections. Look at the layout, formatting, and font. Decide what you like and what you do not like.

Get Your Information into a Workable Form

To start making your plan, you need your information (ideas, calculations, data, etc.) in a form that is easy to arrange and rearrange. Decide on the key points for each section of your thesis or dissertation. You could note the points on paper, on computer, or write them on index cards. Once you have your key points, shuffle them about until you are happy with the order.

Creating a Plan for Your Thesis or Dissertation

One sensible approach is to make notes for your project in the following order:

> **Materials and Methods**, also known as 'Experimental', 'Experimental techniques', 'Sampling strategy and methodology', 'General procedures', 'Data acquisition and processing', etc. This chapter of your thesis covers what you did and how you did it.
> **References**. This chapter tells the reader where you got your ideas and information from.
> **Results.** This states what you learnt from your experiments.
> **Introduction.** This introduces the reader to your field of research, your aims, and the experimental system in which you have been working.
> **Discussion**. This should relate the significance of your findings to your field of study, and give your conclusions and suggestions for future research.
> **Abstract**. This is a condensed version of your whole dissertation or thesis.

Planning your project in this order helps you to see what you did (Materials and Methods) and what you have achieved (Results); which is often different from what you set out to do and achieve. It then helps you to re-evaluate your original aim, and to alter that aim if you have not achieved it, so you can place your results and conclusions in the best context.

Planning Materials and Methods

This chapter of your dissertation or thesis tells the reader what you did and how you did it; it is really just a recipe section. The Materials and Methods that you present must be absolutely accurate because someone reading your thesis or dissertation should be able to repeat your work exactly. Include all the details they might need, such as the pH of solutions, the names of manufacturers of chemicals and apparatus, etc.

Do not simply set out your materials and methods in the order in which you did the experiments. Look through your materials and methods, give each a heading and then group them according to type. Within each group of materials and methods put the generally used materials and methods first, followed by the more specialised ones. If necessary include your methodology for statistical analyses, approximation methods and estimates of error, etc. Do not include all the materials and methods you used—just the ones that are relevant to your final project.

If you have written computer programs that could come under materials or methods it may be easier to put them in an appendix. If you have used available computer programs and databases or Web sites in your research, reference them fully.

For some subjects, particularly theoretical disciplines, it will be necessary to carry out a literature survey before planning this section of your thesis or dissertation (see *Chapter 12: Resources*).

Fig. 1.1 Describe any unusual or specialist materials or method.

Do not confuse materials and methods with results

Your Materials and Methods chapter is simply a set of instructions for the reader—like giving the recipe for making a cake. Results are what you found out from your experiments, the data you have generated. *Chapter 3: Planning and Writing the Materials and Methods / Experimental Techniques* will give you further information on the exact contents and how best to arrange and lay out your work, including the use of appendices and tables.

Planning Results Sections

The results are the core of your thesis. You need to present them well so your examiners can see what you have achieved. In your Results chapter(s) you have to explain why you did your experiments and what you learnt from them. Remember, you are not giving detailed protocols, which are in Materials and Methods (see *Chapter 4: Planning and Writing the Results*).

The first thing you need to do is get your aims clear. Why have you been doing your research? What have you been trying to show in your experiments? Try and pose your aim as a single question or statement. You can then arrange your results to best address this aim. Spend time looking though your notebooks and noting all your results, even for those experiments which went wrong. Keep these notes simple, just one sentence for each main result. Then, before planning your Results chapter in full, do a literature survey—a review of all the literature (books journal articles, Web sites, etc.) that might be relevant to your project. You could leave the literature survey until you write the Introduction, but doing it at this point will help remind you of the significance of your results (see *Chapter 12: Resources*).

Decide on your most important results and the order in which to present them. Start with the results that are the simplest and underpin your other work. Once you have set these down and are on solid ground, move to the next result, building to support your aims. Present your results in the most logical and persuasive order; this might not be the order in which they were produced. Do not present irrelevant experiments and results simply because you have them; they may confuse your examiners and are unnecessary.

Once you have ordered your results into coherent groupings it is worth considering how many Results chapters to have. If you have a number of markedly different key results, form a separate chapter around each of them. Generally there are no rules as to how many Results chapters your thesis should have; divide the Results into as many chapters as are necessary to group your work into logical and easily understood sections.

If you have a number of Results chapters it often makes sense to provide a brief introduction to each one, describing your strategy and specific points relevant to that section, followed by the results themselves and then a short discussion. This will give the reader a detailed critique and an immediate understanding of each result. The wider implications of your findings can be covered in the Discussion chapter.

Figures, tables and appendices are often extremely helpful for summarising a large amount of experimental data. Draft these before creating them and discuss the main points of each one in your text (see *Chapter 7: Figures and Tables* and *Chapter 8: Deciding on a Title, and Planning and Writing the Other Bits*).

You will probably find that as you write your Results chapter you think of points that should go into your Introduction and Discussion chapters. Keep a note of these as you go along, either on a computer file or in a notebook.

Planning Your Introduction

In your Introduction, start broadly laying out the background of your research. Then narrow down to your project and the specific question you are trying to address—your aim. Finish the Introduction with a few points about how you have tackled the question experimentally, to lead the reader into your Results chapter (see *Chapter 5: Planning and Writing the Introduction*).

If you did not carry out a literature survey when planning your Results you must do one now, so that you fully understand the background to your research. Cite your references as you write, so you know each of your statements can be supported by publications.

The beginning

Begin by giving the reader the background to your project. Note down all the points you want to make—just key words and simple sentences. Arrange these points so you start by describing the field in which you are working. Either give a short history of your field of study, including the main theories and findings, or simply review the current situation (which is probably easier). Remember that the reader cannot ask you questions while they read, so you have to provide them with all the information they need to understand your project.

The middle

Next, narrow down to your particular area of interest: tell the reader why it is interesting and why you chose to study it. Cover any important findings or theories which led to your project or which affected your work, referencing wherever necessary.

The end

State your aim and then give a brief introduction to your experimental approach to prepare the reader for your Results chapters.

Introduction and literature survey sections

In some disciplines it is appropriate to divide the Introduction into two chapters, the first describing your project and its significance, the second providing a review of the literature and/or the theoretical background to the project, including mathematical concepts or underlying theories and experimental approaches. Organise your Introduction or introductory chapters into logical and easily understood sections.

Planning Your Discussion

Your approach when planning the points for your Discussion should be the opposite to that when planning the Introduction. Start with the experimental aspects of your findings and consider how your results have addressed your aim. Then broaden out to show their relevance to the general field of your research and state the conclusions that can be derived from your work (see *Chapter 6: Planning and Writing the Discussion*).

The beginning

Start by outlining the general thrust of your argument—restate your aims.

The middle

Discuss your results individually, then relate them to your field of research. Be fully aware of the background to your project, the Introduction, because this may affect your conclusion.

The end

Next you will give your conclusion. Make sure it is supported by your results and discussion. Some disciplines favour putting the conclusion in a separate section, so check to see what is conventional in your field.

At the very end, whatever your results and however successful you have been, finish on a positive note by pointing out interesting avenues for future work that arise from your project. If appropriate, 'suggestions for future research' may be placed in a separate section.

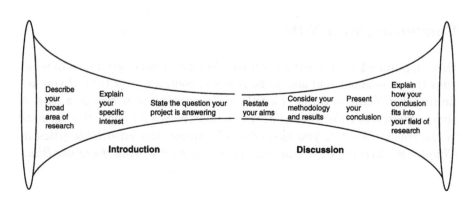

Fig. 1.2 The structure of the Introduction and Discussion.

Planning and Writing Your Abstract

An abstract is a condensed version of your whole dissertation or thesis. Your Abstract should be short, not more than one side of paper, and without headings. Make it a simple, positive and punchy account of your project that addresses the following:

- what question are you asking?
- within which experimental system are you working?
- what are your results?
- what is your answer to the question posed?

As with the main body of your thesis, while this reads logically and is the best order for presentation, it is not the most sensible order in which to write. Plan your Abstract by writing brief statements about your project in this order: (1) Materials and Methods/Experimental Techniques; (2) Results; (3) Introduction; (4) Discussion. Then rearrange these statements so that they answer the four questions in the order above.

For some degrees, such as some doctorates, you are required to submit an Abstract and Title to your university several months before submitting your thesis. Find out if this is the case for you. Writing an abstract is a good way to start you thinking about the contents of your thesis and give you a concise version of the final work, which will help you with later planning (see *Chapter 8: Deciding on a Title, and Planning and Writing the Other Bits*).

Composing Your Title

Having planned your Abstract you should have a fairly good idea of what your thesis is about and should be ready to compose your title. If you are finding it difficult to think of a title, try re-stating your aim as a title. A short descriptive title is best, one which tells the reader about the contents of the thesis. We give you examples of suitable and unsuitable titles in *Chapter 8: Deciding on a Title, and Planning and Writing the Other Bits*.

Figures, Tables and Appendices

Figures, Tables and Appendices are used for presenting your data to the examiners and for explanatory diagrams of something you are discussing in the text. Plan their contents carefully. Discuss only their most important points in your text, you do not need to discuss every detail. Each figure (including graphs), table and appendix needs a title, and figures and tables also need annotation and legends. Each one must also be referred to in the text. Put figures and tables as close as possible to where you discuss them; appendices usually go at the back of the thesis.

Prepare your figures before you print the final version of your thesis or dissertation. If you number each figure in advance it will help you spot any errors, such as having two figures with the same number (see *Chapter 7: Figures and Tables, Chapter 8: Deciding on a Title, and Planning and Writing the Other Bits*, and *Chapter 9: Proofreading, Printing, Binding and Submission*).

Writing Throughout the Course of Your Project

If you begin writing your thesis on computer as soon as you start your project, you will have less work to do at the end. Begin with your Materials and Methods/Experimental Techniques and References.

Materials and methods/experimental techniques

Write your Materials and Methods into a word processing file as you go along. Do not worry about organisation of this information at this point. Writing your materials and methods now will help ensure they are accurate and that you have included all of them. This is a tedious section to write if you leave it all until the end, when you could well be panicking or bored of the whole thing, you could easily make mistakes and omissions (see *Chapter 3: Planning and Writing Materials and Methods/Experimental Techniques*).

References/bibliography

When you come across useful references, enter the full reference description into a reference database program or word processing file. This description generally includes: a list of the authors, year of publication, title, correct journal abbreviation, volume number and page numbers. There may be regulations as to what reference details your department or university requires in your thesis, so make sure you know these before you start adding your references (see *Chapter 2: Planning and Writing the References (Bibliography)*).

Start to collect a library of photocopies of important references, so that you have these to hand when you need them.

Overall Content

Very roughly, your Introduction will be 20–30% of the whole, your Materials and Methods/Experimental Techniques 10–20%, your Results 35–45%, and your Discussion 20–25% (not counting your Title Page, Abstract, Table of Contents and other appendices).

Writing

When you have your detailed plan, put all your notes together in the correct order and check that they make sense. Show the complete thesis plan to your supervisor and make whatever changes they suggest. You can then begin writing.

When you open your first word processing file, make sure you have the correct margins and (as far as possible) the correct page layout, so you do not have to make time-consuming changes to layout or the figures when you come to print the final draft. Usually you should use double or one and half line spacing, make sure you have sufficient margins around the text; leave at least 4 cm in the left margin for binding the pages together. Find out if your department or university has any rules about the format and layout of the thesis or dissertation. If there is a word limit for your thesis, remember to use your word processing program to carry out a word count regularly so that you can keep track of the size of the thesis (see *Chapter 13: Layout*).

Make the figures and figure legends for each chapter as you go along; making them can be a welcome relief from typing, and you do not want to have to do them all at the end when you will be running short of time (see *Chapter 7: Figures and Tables*).

Putting Down the First Words

Before you start writing have your plan in front of you. Use it when you write—that is what it is there for. Some people like to start by getting something down on paper, whatever it is. They then use these rough jottings as a basis from which to write their text. Others prefer to spend much longer on the initial writing, getting it as near perfect as they can. Choose whichever approach suits you best. If you are using the 'put anything down and change later' method, do not be too free in your approach to the first draft: do not go for stream of consciousness writing—it can be a nightmare trying to sort out your ideas later.

Drafts

Your text will inevitably go through a number of drafts, pruning and clarifying the language each time, before it is ready for submission. With a large text such as a thesis or a dissertation you will probably go through this drafting process for each of your chapters or sections individually—some will need more re-drafting than others. When you complete a draft print it out on paper—you will get a better idea of how it looks, and it is easier to spot mistakes on paper than on screen.

It is best to give yourself a break between writing and checking your drafts. It is a lot easier to spot mistakes when you have put a bit of distance between you and the draft. Try and imagine yourself as another scientist to see if your text makes sense. If you can persuade a friend or colleague to check it for you this will help, as they will spot mistakes that you have missed. Make sure they know at least a bit about your subject and have a good command of English.

Give a copy of each draft chapter to your supervisor to read and correct. Make the draft as good as you can so your supervisor will not waste their time correcting small mistakes such as spelling errors. Take note of your supervisor's comments and corrections (see *Chapter 11: Supervision*). Go through the draft adding information you have missed, correcting mistakes and awkward constructions.

When your thesis or dissertation is in its final stages, add the following sections (not all of which are necessary for every dissertation or thesis, see *Chapter 8: Deciding on a Title, and Planning and Writing the Other Bits*):

Title Page
Abstract
Dedication
Acknowledgements
Table of Contents
List of Figures
List of Tables
List of Appendices
List of Abbreviations (also known as 'Nomenclature' in some disciplines)
References (also known as 'Bibliography' in some disciplines)
Glossary
Appendices
Published Papers

Before you finally submit your thesis make sure it is thoroughly and accurately proofread.

Use of English

Your text is a serious scientific document and it should be prepared with the same attention to detail as would be expected of a paper submitted to 'Nature' or 'Science'. Reading a dissertation or thesis can be as hard work for an examiner as writing one is for you, so your text has to be presented as clearly as possible, without ambiguity, and with sufficient introduction to be understandable. Examiners have very little time between teaching, grant writing, administrative duties, etc. and they may well have to read your text on the train, or in between taking the kids to ballet practise and trying to get the exhaust fixed on their car. Make the job of reading your dissertation or thesis as easy as possible: the harder an examiner finds your text to read, the less well inclined they will be towards you and your project—which is not a good thing. The harder you work on making your text clear and easily understandable, the less work the examiner will have to do deciphering your text and the better disposed they will be towards you and your project—which is a good thing. Keep your style crisp and to the point, be concise, and use plain English. Use accepted scientific terms wherever they are appropriate. Use the correct units and their abbreviations. Use headings, figures and tables to break up your text into easily readable sections so that both you and the reader can keep track of it. In your Introduction and Discussion you can be more discursive than in your Materials and Methods, and Results, but keep your writing concise and focused all the same. Use appendices for information that is a necessary part of your project but would clutter your text if put in with it.

Avoid wordiness, vagueness, colloquialisms and contractions (*it's, lab,* etc.) and understand specialist terminology if you use it. Use the spell-checker on your word processing program; there is no excuse for spelling mistakes. Check your grammar is correct and appropriate. It is best to avoid the passive, 'the digests were continued', it can give you the right tone of detachment, but can also be very wordy; try and use active constructions wherever possible, for example, 'digestion continued' (see *Chapter 14: Use of English* and *Appendices: 1, 2, 3,* and *4*).

The Importance of Good Presentation

When people go for job interviews they make an effort to look smart—which shows that they are taking the interview seriously. If your manuscript looks good the examiners are more likely to take you and the project seriously. Take into account any conventions of your discipline or department and stick to the same format and layout consistently throughout your thesis or dissertation. Before you submit your thesis print a complete copy, check and proofread it thoroughly one last time (see *Chapter 13: Layout* and *Chapter 9: Proofreading, Printing, Binding and Submission*).

Plagiarism

Plagiarism is stealing other people's ideas, writings, or inventions and passing them off as one's own. When using other people's ideas or writings, make it clear that they are the 'property' of their author by showing that you are quoting and giving a precise reference. If a student plagiarises someone's work and the theft is discovered, which it almost certainly will be, they can be failed without further question.

Binding

If you are binding your papers together yourself, organise what you need in advance, and ask your supervisor for advice on what your department prefers. Some theses and dissertations, such as doctoral theses, need to be professionally bound. Find out in advance where the binders are, what they charge and how long they take.

Fig. 1.3   Check when the binders are open.

Where to Write

Writing your dissertation or thesis will be hard enough at the best of times. Find somewhere to write where you can concentrate on your work and will not be interrupted too often. You will be spending a lot of time in this place so make it comfortable. Your concentration will not be helped if you develop chronic backache, so check your chair is the right size and the table or desk the right height for you. Make sure you have easy access to cups of tea or coffee, biscuits, etc.

Resources

Notebooks

Organise your information well and keep it safe. Some of your information, such as notes from experiments, will be irreplaceable, some will take a lot of time to find again. Keep a small notebook or pad for jotting down ideas that could come to you at any time, in bed, on the bus or while eating breakfast.

Libraries

All libraries offer the same basic services: providing books and journals, ordering articles for you, and allowing you to carry out reference database searches. They can also supply you with information such as lists of citation abbreviations and SI units. Get to know your librarian, who will be able to show you how the library and its facilities work and point you in the right direction when you are looking for information (see *Chapter 12: Resources*).

Your computer

Learn to understand and love your computer, whether you have your own or are using one in your department. Your computer not only works as a magic typewriter that allows you to shift chunks of text around and correct spelling mistakes at the click of a mouse; it also allows you to sort, store, and retrieve your information easily. Take time to learn how to work with the programs you are using (word processing, graphics, reference database programs, etc.) and peripheral devices such as scanners (see *Chapter 12: Resources*).

Using the Internet and World Wide Web

The Internet and World Wide Web can provide you with a lot of useful information, from important references to moral support from other people writing dissertations and theses. Learn how to use and enjoy them (see *Chapter 12: Resources*).

A reminder about good housekeeping

Remember to keep copies of all your computer files on at least three floppy disks, and at least one on a hard disk. As disk to disk copying is where most problems occur, it is not enough to have a copy of your valuable thesis files on a disk with just one backup floppy disk. If you have not backed up your thesis and lose it, you will have no excuses. Also keep up-to-date dated printouts of all your documents just in case you lose everything you have on computer (see *Chapter 12: Resources*).

Your most important resource is yourself

Remember, you are a human being and human beings do not operate like machines—although you might well feel like one at times. Obviously, you will have to work hard on your text but do not push yourself to the limit every day. Learn to manage your time and your tasks, alternating between boring repetitive jobs and more interesting ones. Set yourself achievable deadlines for each piece of work but do not get too upset if you run over your deadline, as inevitably you will from time to time.

Fig. 1.4   The essentials of thesis writing.

Interim Reports

You may well have to submit interim reports for your project, for example, quarterly or first year reports. Approach them in the same way as your dissertation or thesis, although they will mainly be concerned with materials and methods, and results. Try to produce as professional and well-written a document as possible. These reports might well be useable as parts of your final dissertation or thesis, which will save you a lot of work at the end of your project.

We will cover each chapter of your dissertation or thesis in detail in later chapters of this book and provide you with practical suggestions for presentation of data. We have also included a number of useful appendices.

Remember: start from a well thought-out detailed plan and you will succeed!

Key Points

- Ask your supervisor to recommend a good recent dissertation or thesis in your own field as a guide to form, content and style
- Read your university or departmental rules on thesis submission, before you begin writing

- Make a detailed plan of all your sections and chapters before you start writing; review your work in the order: Materials and Methods/ Experimental Techniques, Results, Introduction, Discussion
- Find out if there is a thesis or dissertation template for your department, and if so, use it
- Type Materials and Methods/Experimental Techniques, and References into appropriate computer files throughout the course of your project

Chapter 2

PLANNING AND WRITING THE REFERENCES (BIBLIOGRAPHY)

To avoid confusion, we use *to cite* to mean to refer to a piece of work in your text, and a *citation* or *citation mark* to mean the indication in the text that you are doing so:

> The plumage on most south African landfish is yellow **(9)**, but specimens with blue or red plumage are occasionally seen **(10,11)**.

> The plumage on most south African landfish is yellow **(Michael and Ridgeley, 1996)**, but specimens with blue or red plumage are occasionally seen **(Collins *et al.*, 1997, Gabriel *et al.*, 1998)**.

By *reference* we mean either the original document or the directions for the reader to find the original document:

> 9. Michael, G. and Ridgeley, A. (1996). Survey of South African landfish colouration. Landfish Today 4: 5–23.

Why Reference?

Any piece of scientific writing, whether it is a journal article that presents important new results or a first year essay, has to be properly referenced because the reader must be able to check the original sources of any ideas or findings that you have used.

What to Reference

Strictly speaking, you should reference any information you have used that is in the public domain—that can be found and read by your readers; this includes other dissertations and theses, your own published work, and conference proceedings. In practice you do not need to reference statements of general knowledge ('the earth is round') but you do need to reference information which is not yet generally accepted or known ('core samples from the University of West Cheam contain extra-terrestrial rocks (McInroy *et al.*, 1998)'). If you have used any information from your own publications then remember to reference these as well. You should also reference methods you have used, including computer programs.

What to Cite

Give a citation in your text every time you make use of a reference. You can cite information for which no written reference exists, for example a conversation or seminar talk, but avoid this if possible because the reader has no way of checking your information. If you cite work that is not in the public domain make sure you have any appropriate permission before doing so.

How Many References and How Old?

There are no rules as to how many references a thesis or dissertation should contain, as long as the reader can see where your information comes from. A first year report may cite only 20 or so references, while at the upper end of the scale, a well referenced life sciences PhD thesis can cite two or three hundred papers; ask your supervisor to check that you have enough references.

Science is very fast moving and you should ensure that your references are up-to-date. If your most recent reference is five years old this is a clear sign to the examiners that you have not been keeping up with your field and are unlikely to have a good background knowledge, or even to be particularly interested in your research. Make a point of checking you have included the most up-to-date relevant references you can.

Problem References

Sometimes it is impossibly difficult to track down an original document, perhaps because it is very old or in an extremely obscure foreign journal. In this case, indicate you have not read the original but know that it is useful because you have seen it cited in other papers, by writing *cited by* or *cited in* with your reference. However, be very cautious of using other people's references in this way as you are putting a lot of faith in their judgement and thoroughness.

Quoting

If you need to quote the exact words of the original statement make sure you get them right. If you are using short quotations, less than about 20 words or so, you can simply put them in inverted commas in your main text (see the punctuation section in *Chapter 14: Use of English*); with larger sections of text it is better to put them in a 'block indent', with the citation in the text underneath, like this:

> Heat generation from normal bacterial metabolism remains a difficulty for researchers in this field. Many attempts have been made to overcome this problem, Karloff maintains:
>
>> The normal phases of catabolism, in which complex substances are decomposed into simple ones, and anabolism, in which complex substances are built up from simple ones, have been bypassed by the technique of Moog Genesis.
>>
>> (Karloff, 1996)
>
> These claims are highly suspect and many other studies show that such measures of energy function may be artefactual and arise from the unusual measurements taken by Dr Karloff at the University of West Cheam.

The Implications of Copyright

There are strict rules about photocopying and quoting other people's work (see *Chapter 12: Resources*).

Collecting and Storing Your References

Build your own library

It is a good idea to build a personal library of the original papers you are using as references, so you will have them to hand when you need them. We look at this in more detail, including copyright restrictions, in *Chapter 12: Resources.*

Fig. 2.1 Build your own library of references.

Enter the reference into a reference storage file

When you come across a useful reference, write the reference details (the name of author, year published, etc.) into some kind of computer file. Using a reference database program is the easiest way of storing and retrieving your references (see *Chapter 12: Resources*). If you do not have access to a reference database program then type your reference details into a normal word processing file.

Start now!

Your references are the single most tedious item you will have to type into your thesis. If, from the start of your project, you enter references into a database as you come across them, you will be able to insert them into your text as you write, without having to track them down and type them every time you need to cite one. If you have not been doing this, start entering your references into some form of computer file, right now! It will get you back in contact with parts of research you might have forgotten, and will save you having to do this painstaking and error-prone task right at the end of the thesis writing process when your time could be much better spent honing the more interesting and creative aspects of your writing.

Add Citations to Your Text As You Write

When you make a statement that needs referencing, put the citation in your text as you go along rather than writing your text and adding citations later; that way you will be able to make sure any information you are presenting has a reference that goes with it. Citations are usually placed at the end of the sentence so that the flow of the sentence is maintained.

Statements or methods that are presented in theses without being referenced imply the writer may not have a good background understanding of their field, or, far worse, made it up as they went along. If you cannot find an appropriate reference to cite it may mean that you need to do some more reading around the subject and explore a little more deeply before you continue writing. Choose your references with care, and make sure you really have read them; by citing a paper you are saying '*I have read this paper*'.

The Final Printout

When you have the final version of your thesis, carry out a check for references, especially if you are not using a reference database program, but are adding references from a simple word processing file. Go through the text and whenever you see a citation, tick off the corresponding reference in your References section, so you do not cite papers that are not included in your Reference list, and that you do not have papers in the Reference list which are not cited in the text.

Formats for Citations and References

Scientific journals have strict rules as to how papers are presented, including how the references are cited. Check if there are any regulations concerning the format of your references; London University, for example, requires references for theses submitted by MD, MS, MDS and D Vet Med students to be organised according to the Harvard system (see below). If there are no specific rules for your thesis or dissertation it is worth finding out if there are any conventions in your field or in your department. If there are, you would be sensible to follow them—ask your supervisor and have a look at a good, recent thesis in your field.

The Harvard System

With the Harvard System the names of the authors and the date of publication are included in brackets in the text:

> ... coulomb explosion imaging is of limited use in this field (Simpson and Vaidynathan, 1994) and so was not used...

This system can get a little unwieldy, and it is best to avoid citing too many references in one block, but it reminds both you and the reader which papers are being referred to. Where the name of the author occurs naturally in the text you only have to add the date in brackets:

... Halliwell's assertion (1998) that partons, such as quarks, are actually malevolent fairies is extremely controversial...

... as Harrison and Davies (1994) state, the β-form of lithium sulphate is made up of molecules in an hexagonal array...

If there is more than one author cite both names on two author papers. For papers with three or more authors, cite the first name only, followed by the words *et al.*, from the Latin, *et alii* meaning *and others*. Normally *et al.* is italicised because it is a Latin phrase (see *Chapter 14: Use of English* and *Appendix 5 Latin Words and Abbreviations and What They Mean*):

... if a pair of opposed and external forces is applied to a body, the relative positions of this body's components change (Jagger, 1994, Watts *et al.*, 1996, Richards and Wyman, 1997), and new restoring forces arise (Mercury *et al.*, 1998)...

You will often have to refer to more than one paper, in which case it is usual to cite them according to the order in which they were published—starting with the oldest paper (Jagger, 1994, Watts *et al.* 1996, Richards and Wyman, 1997). When you cite papers from the same year, put them in alphabetical order according to the name of the first authors.

Whilst you would not normally include initials it makes sense to do so if you are citing two or more papers with the same surname and date (for example, Sinatra, F., 1969; Sinatra, N., 1969). When citing two or more papers with the same first author in the same year, indicate this by adding *a, b, c,* etc., after the name or year number. For example, Guthrie, A. (1997a); Guthrie, A. (1997b).

With the Harvard System, the references should be arranged in alphabetical order in your References chapter, according to the name of the first author. If you have many papers with the same first author, arrange them with the earliest first, and most recent last.

Numerical system

Some institutions and disciplines prefer the numerical system:

> ... if a pair of opposed and external forces is applied to a body, the relative positions of this body's components change (5–7), and new restoring forces arise (8)...

This does not disrupt the flow of the writing as much as the Harvard system, but the reader has to look in the Reference chapter to see what the number relates to.

If you are using the numerical system you can put your citations in brackets:

> ... Students at West Cheam University have undertaken interesting research that shows the time-space continuum is distorted along the axis of a ley line running between Nonsuch Park and Uppingham (9). Dr Karloff, Dean of Alternative Physics at the University, disappeared for 30 minutes during a seminar while demonstrating the effect of Moog Genesis in the vicinity of this ley line (10, 11).

or some disciplines prefer to put citation numbers in superscript:

> ... A recent study[12] has revealed disturbing evidence of fluctuation among time zones in South Kensington...

When you cite more than one paper at the same point in the text, cite the oldest one first. Papers from the same year are cited in alphabetical order.

Generally when using the numerical system the reference details are arranged in numerical order in your References chapter; in a few subjects it is acceptable to place the reference in a numbered footnote at the bottom of the page containing the citation.

How to Write the Full Length Reference

There is a minimum amount of information that you should give to ensure the reader can find the original reference. For example, for journal articles you need to include: the name of the authors, the name of the publication, the volume number, and the first page number of the article or chapter. Try to make things easy for the reader so include as much information as possible.

References from journals

Here are a few examples of possible layouts:

> Kravitz L. Roachford A. 1998. The Life Cycle of Landfish in South Coast Resort Towns. Landfish Today. **4:** 8–12.

> Kravitz, L. and Roachford, A. (1998). The life cycle of landfish in south coast resort towns. Landf. Tod. *4* 8–12.

> Kravitz, L. Roachford, A. (1998). The life cycle of landfish in south coast resort towns. *Landf. Tod.* 4 8–12.

Make sure you include the following:

Surname of all the authors and initials

It is best to cite the authors in full unless this would be very unwieldy (if for example a paper has 40 or 50 authors). *Prince, A.F.K., et al.* does the job, but is a little unfair on the other authors (one of whom might be your examiner). It is best to include the initials, particularly with common names, such as Fisher, Wong or Patel, to make it clear exactly which Fisher, Wong or Patel the author is.

If the article has been published by a team or institution rather than individual authors, list the group by name:

> West Cheam Landfish Taskforce (1996). Landfish Survey: West Cheam and Adjacent Areas. *Landf. Tod.* 2: 43–58.

Year of publication

This is essential. Remember, it is best to use the most up-to-date references as possible, because this shows the examiner that you are following developments in your field.

Title of article or paper

This tells the reader about the contents of the paper—it can also be a useful last minute reminder before an oral examination. Give all titles in their original language, with a translation in brackets if necessary.

Journal name
The journal name is the only way people can track down the reference, so you must include it. People usually do not bother to write the journal name in full because every journal name has a standard abbreviation, which you should use. If you are not sure of a particular abbreviation you can usually find out by checking the journal in question. If you have difficulty ensuring you have the correct abbreviation the following reference book, which is probably held in your main library, lists science journal abbreviations:

> *Alkire, L. Periodical title abbreviations. 10th edition. 1996. (Gale Research, Inc., Detroit).*

Volume number
This is essential for allowing the reader to find the article. You do not need to include the issue number for each volume.

First and last page number of the article
In order to find the article quickly the reader needs the first page number. Giving the last page number will make the examiner more inclined to believe that you at least looked at the article and did not just copy the reference citation from another paper. Check that your last page number is always greater than your first page number.

References from books

Follow the same principles when writing the reference details for books. When you reference books, cite the authors, the date, title, the town or city where it was published, and the name of the publisher. For example:

> Bowie, D. and Mitchell, J. (1994) A logic-based calculus of everyday objects. (Hendrix and Joplin Press, San Francisco).

If you are citing a chapter from a book then include the above details, plus the name of the chapter, and the chapter authors, as well as the book title. For example:

Downward, J. and Tybulewicz, V. (1998) Successful research methods. In *Applications of Signalling*. (Nottinghill Press, London)

If the book is edited then include the names of the editors in your reference details:

Bopper, B. and Miller, G. (1997) Statistics don't mean nothin'. In *A Handbook of Paranormal Mathematics*, Katsanis, N. and Bermingham, N. (Eds) (West Cheam University Press, West Cheam)

You can also include the page numbers if you want to refer to just one section; *p* is used to refer to one page and *pp* to refer to a number of pages.

Ashworth, A. and Marshall, C. (1997) Probability theory and gene cloning. In *Favourite Oncogene Stories*. (Fulham Publications, Cambridge), pp. 103–128

Sourounian, L., Doering, D. and Swendeman, S. (1998) The growth of red giants. In *Stellar Evolutionary Theories*. (Penn and Rockefellar, Boston), p. 206

Also list the volume number if there is one, and edition number if it is necessary for the reader to find the reference:

Karloff, B., Greenwood, B. (Eds). 1963. Revitalisation of the Dead. Vol. 2. (Head and Throttle, Whitby)

If the publisher lists several cities—for example, *London, New York, Toronto, Sidney, Singapore*—you only need to list the first one.

Theses and dissertations

These are referenced in much the same way as books:

Morissette, A. (1999) The effect of sharp-edged prophylactics on patient recovery. (PhD Thesis). University of West Cheam. p. 963.

Conference abstracts and proceedings

Published abstracts and proceedings from conferences are referenced similarly, including the name of the conference and where it took place:

> Bentliff, G. and O'Donovan, T. (1998) Diffusion artefacts of scanning tunnelling electron microscopy. Fifth International Workshop of Electron Microscopic Techniques. Humbolt, Canada.
>
> Garling, H., Garling, O. and Slattery Z. (1997) Disposal of the products of fermentation. Second European Conference on Catabolytes. Rogera, Spain.

Some conference proceedings state that you are not allowed to quote material from the abstracts without permission of the author(s).

Unpublished papers and talks

Give the speaker's name, date, title of the meeting, place of meeting:

> Dog, S. D. Ballistics in a Modern Context. Paper read at conference on Psychology of Risk. (University of West Cheam, June, 1998).

If you are giving information that someone has told you and it is unpublished and cannot be referenced in any other way, then put 'personal communication', or 'pers. comm.' as a reference—but avoid doing this if possible as the reader cannot easily check your source.

Computer programs and the World Wide Web

If you have used available computer programs, reference them fully: give the version number and reference the original paper in which the program appeared (or the name and address of the supplier of the program). If you have been searching databases then give either the version number of the database, or if the database is frequently updated give the date of your final search. If you have been using data or programs from World Wide Web sites give the name and address (http://www...) of each site.

The Importance of Referencing Carefully

Referencing properly, although tedious, is not difficult. Never forget that your examiners are very very human and that one of yours might have a particular bee in their bonnet about sloppy references. They could well have written their own thesis in the days before such fancy modern aids as reference programs, and might bear a grudge as they recall the hours they spent typing their thesis references when all their flatmates were out at a Sex Pistols concert.

Common Mistakes

- Citing a paper in the text, and forgetting to include it in the References/ Bibliography section. Even with the use of reference database programs, losing references from the text is still a common sin. Avoid this by scanning through the final version of your thesis and as you come to each citation, tick off the appropriate reference in your Reference chapter
- Using incorrect journal abbreviations. There is no excuse for this because it is so easy to find the correct abbreviation. Examiners can, and do, ask for corrections to errors in journal names, and this can be very time-consuming. See:

    Alkire, L. Periodical title abbreviations. 10th edition. 1996. (Detroit, Gale Research, Inc.)

- Inconsistency in the style of citations and references. For example, these three references appear in the same Reference section, and each one is irritatingly different from the other:

    King BB and Wonder S (1996). An integrated protocol for locomotor and cognitive testing of landfish. *Landf. Tod.* 2: 8–12

    McCartney, P., Starr, R., Lennon, J., Harrison, G. (1998). The biology of northern landfish. *Landfish Tod.* 4 21–27.

    Zappa *et al.* (1997). *Landf. Today* 3 21.

- Mis-spelling authors' names (especially on well-known papers). For obvious reasons, mis-spelling an author's name will count against you even more than normal if the author happens to be one of your examiners *A friend of ours of Eastern European origin, with a long but not particularly difficult to spell name was asked to examine a PhD thesis. This examiner found his name cited throughout the text of the thesis— and misspelt every time, in a variety of ways. During the viva the student was made to suffer for his mistake and had to spend several days correcting all the offending citations.*
- Getting the page numbers wrong. Guard against this by carrying out a quick scan of your references and check the last page number is higher than the first page number
- Getting really famous references, whose details the examiners know, wrong; for example, citing the discovery of the structure of DNA in a paper by Wilson and Crick

Key Points

- Write your references into a computer file as your project proceeds, preferably a reference database file of some sort
- Do not be lazy with references; statements made out of the blue, without the support of a reference, are obvious to experienced examiners
- Take care with references. It is very easy to make mistakes when typing them and examiners know this
- Use a consistent style for citations and references
- For journal articles enter authors, year, title, journal with correct abbreviation, volume number, page numbers, according to the conventions of your field
- For books enter all authors, year, title, editors (if there are any), publisher and the town or city in which they are based
- Make sure your references are up-to-date

Chapter 3

PLANNING AND WRITING MATERIALS AND METHODS/EXPERIMENTAL TECHNIQUES

The aim of your Materials and Methods/Experimental Techniques sections(s) is to produce a detailed recipe section for your experiments. By 'Materials' we mean the things you use for your experiments: reagents, theories and theoretical models, equations, human or animal subjects, equipment, existing data you are analysing, etc. By 'Methods' we mean the processes you use.

Most disciplines, whether practical or theoretical, will have a section or chapter that fulfils this function, whatever the section is actually called: 'Mathematical Concepts', 'Techniques', 'Experimental', 'Experimental Techniques', 'Sampling Strategy and Methodology', 'General Procedures', 'Data Acquisition and Processing', etc. The conventions of this section vary widely between different disciplines and sciences, for example, some chemists might not include a Materials section, whereas ecologists might separate this section into Processes rather than Materials and Methods. Whatever approach is normal in your field, the underlying principles are the same.

Some theses, such as those written for certain medical degrees or theoretical subjects—for example, cosmology or mathematics—may consist of a number of published papers sandwiched between an overall introduction and conclusion. In these cases, the Materials and Methods sections, whatever they might be called, will be included within each paper. Look at a good recent thesis from your department for your local customs.

Whatever the Materials and Methods/Experimental Techniques section is called in your discipline, and however it is arranged (whether as one chapter in your thesis, or a section in each of the papers you are including in your thesis), your materials and methods have to be clearly and completely stated so that other people can use your exact methodology, including precisely the same materials (whether they are chemicals or theoretical equations) to replicate your results. You can assume some prior knowledge on behalf of the reader, for example, they will understand standard scientific terms within your field. If in doubt about how much prior knowledge to assume, err on the side of caution—ask your supervisor for advice and look at a good recent thesis or dissertation.

The Importance of Forward Planning

As we said in Chapter 1, start writing your Materials and Methods into a word processing file, at least in rough form, as soon as possible—preferably from the beginning of your project. If you do so you will have far less work to do at the end and will not be in danger of losing or forgetting important details of your project.

If you have all your Materials and Methods in some form of computer file when you sit down to write the body of your text, celebrate! All you have to do is a little careful editing, which will help you get a good idea of the contents of your thesis, and then you have finished one chapter.

If your materials and methods are only written in notebooks and on various pieces of paper, then start putting them into a word processed file *right now*. Do not leave this job until last, when you will be most likely to make mistakes and might be starting to panic.

Planning Materials and Methods/Experimental Techniques

What to include

Include details of every experiment for which you are giving results; be careful not to overlook minor experiments that are useful for supporting your main data. Include Materials and Methods for experiments that were unsuccessful or contradict your argument if they help you discuss your methodology and make some relevant comments about the main bulk of your results. Unsuccessful experiments should not count against you, as long as you can show that you have learnt from them—they may also be a useful warning to people not to spend time trying similar approaches.

Do not include Materials and Methods for every experiment you carried out during your project. If experiments are irrelevant to your final results they should not be kept in your thesis and will only clutter your text and make it appear as if you do not understand the significance of your own data.

A common error: confusing Materials and Methods with Results

One of the biggest difficulties for people who have to write Materials and Methods sections is confusing their contents with the Results sections. Materials and methods are simply a set of instructions for the reader. Results are what you found out from your experiments, the data that you have generated.

Which to Write First, Materials or Methods?

Some disciplines favour presenting Materials followed by Methods, others reverse this order; have a look at a good recent thesis in your field to find out the local conventions. In whichever order you present these sections it is generally easier to start by planning and writing your Methods and to note down your Materials as they crop up.

Materials and Methods should be presented in a logical order within the individual sections, for example, giving generally used materials, then specialist materials, followed by generally used methods, then specialised methods.

Literature Review

In some, particularly theoretical disciplines, you may need to carry out a literature review (finding and reading papers related to your subject) before you can plan and write your Materials and Methods. During the literature review you will track down, for example, the source of complex proofs or other equations that you need to process your ideas.

Writing Methods

Write each method as if you were simply telling someone what you did. Be clear and do not get too wordy. Recently we saw a thesis in which a student had written '... *the weighing out of the agarose was undertaken until 5 g were measured out and the agarose was then later added to the solution*'; the student could have more clearly and concisely written: '... *5 g agarose were added to the solution*'.

You do not have to reinvent the wheel; but, particularly with more specialised techniques, go into as much detail as a reader needs to be able to repeat the process you used exactly. The small variations in methodologies between laboratories are often important.

Do not explain too much about *why* you have used a certain method. You can put these explanations into your Results section.

Here are some, hypothetical, methods from different disciplines:

2.3.1 Preparation of 1-(*tert*-butyldimethylsilyloxy)-3-methyl-3-(phenyl-sulfonyl) propan-1-ol

Pyridine (5.2 ml, 55.7 mmol, 1.5 eq) was added to a solution of hydroxysulfone (11.3 g, 18.7 mmol, 2.5 eq) in CH_2Cl_2 (70.7 ml) under neon at 4°C. After stirring for 50 mins, TSB (12.7 ml, 27.8 mmol, 2.5 eq) was added dropwise. After 30 mins $CuSO_2$ (aq) was added to quench the reaction.

2.3.1 Isaac-Moss condition

Let $f(x, y)$ be defined and continuous for all (x, y) in the region E defined by

$$a < x < b, \ -\infty < y_t < \infty, \ t = 1, \ 2, \ ..., \ n$$

where a and b are finite, $y = [y_1, y_2, ..., y_n]^T$

and let there exist a constant V such that

$$\left| f(x, y) - f(x, y^*) \right| < V \left| y - y^* \right|$$

holds for every (x, y), (x, y^*) in E. Then for every y_0 in R^n there exists a unique solution $y(x)$ of the problem

$$y'(x) = f(x, y(x)) \quad y(a) = y_0, \quad a < x < b, \quad f: R \, X \, R^n - R^n$$

where $y(x)$ is continuous and differentiable.

2.3.1 Preparation of site for seeding

The field site was located at the half acre field (University of West Cheam, Nonsuch Park, Surrey). An area (40.3×15.0 m) previously sown to grass was ploughed and fertilised (Hodson-Hooker Fertiliser) in mid-August 1996. The site was sown on 23rd April 1998 with a spring crop (*Campanula cochlearifolia*) and netted to prevent theft.

2.3.1 Veronica-Fisher extrapolation

By this method the principal error functions of a given (low order) numerical method are approximated, thereby accelerating its convergence. Suitable linear combinations of approximate solutions obtained by using the low order method on different (uniform) meshes are calculated to complete this process. The Veronica-Fisher extrapolation depends on the knowledge of the powers of j appearing in the error expression for the low order method.

2.3.1 Incubation of adults

Young adults collected from stock sub-cultures were placed in specimen tubes containing cotton wool at the base. Forty healthy adults were selected, covered with yeast extract (Hoyle extract, Whitehead Industries) and placed in each tube. The tube was then sealed with light aluminium foil. Tubes were placed in an incubator and kept at 37°C and 85% relative humidity for 24 hours.

2.3.1 Use of core samples

Conventional core samples from well bore 2 (Salisbury Avenue site, depth intervals: 233.87 to 243.23 m, 274.04 to 285.46 m) and well bore 3 (Cecil Road site, depth interval: 222.00 to 229.17 m) were taken. The samples were photographed and logged; records are given in Appendix 2. Samples were taken at intervals of 1 m.

2.3.1 The Pilli-Vasso first order action

The normal Nicholson-Witherden action was taken as the model, but the metric and affine connections were considered as independent variables. The action:

$$S[g,\Gamma] = \int d^4x (/g/)^{1/2} g^{cd} R_{cd}[\Gamma]$$

was taken where Γ_{cd} is the affine connection. Then by varying Γ it follows that Γ_{cd} is the metric affine connection defined by g_{cd} and by varying g we obtain the Nicholson-Witherden field equations.

2.3.1 Transformation of bacteria

1 µl plasmid (1 ng/µl) containing insert DNA was mixed gently with competent cells and incubated at 0°C for 30 min. The cells were then diluted with 2 ml SC medium at room temperature and incubated at 37°C for 45 min, shaking at 250 rpm. After incubation 100 µl each transformation were plated onto LB agar plates containing 100 µg/ml ampicillin.

Alternatively, *E.coli* Electro-shocker Transformation apparatus (Bioshock Inc.) was used for electro-transformation of cells according to manufacturer's instructions.

2.3.1 Analysis procedure

A model equation, which best fitted the empirical data, was chosen and then a procedure to determine the coefficients of the equation was followed. Regression analysis was chosen in this study. Because of the likely log-normal distribution for x (the weak interaction parameter) the regression analysis was carried out in the natural logarithm of x.

Writing the Materials Section

The 'Materials' section is simply a list of the reagents or equipment you have used in your experiments. This section may have different titles depending on your discipline. Most supervisors prefer you to start with commonly used materials and then list more unusual materials or equipment used for only particular types of experiments. Alternatively, you can simply list all your Materials alphabetically.

As well as the materials and equipment, list your suppliers and any model or version numbers. For some items the products from different suppliers are more or less effective; therefore, someone attempting to repeat your experiments needs to know exactly what you have used. When listing your supplier, it is a good idea to include their head office address otherwise someone, particularly if they are working in another country, might not know how to contact them. Once this address has been cited, you do not need to refer to it again.

As with everything else, different fields have slightly different conventions. For example, a biology thesis may list the suppliers of even the most routine chemicals, whereas a chemistry thesis probably would not. The biologist uses the chemicals as a means to an end and relies on the supplier for quality control, but the chemist may well synthesise, purify and analyse their own chemicals. Ask your supervisor and check a good recent thesis in your subject if you are not sure what to list.

In the example below we show a typical materials section from a molecular genetics thesis:

Chapter 2 Materials and Methods
2.1 Materials
2.1.1 General Reagents
All laboratory chemicals were Analar grade from Springsteen Chemicals (Newcastle, UK) with the following exceptions: Trisma base, ethidium bromide and dextran sulphate were from Cobain Inc. (Seattle, WA, USA); IPTG and X-Gal were from Garcia Industries (San Francisco, CA, USA); sodium chloride, trisodium citrate and potassium acetate were from Cocker plc (Sheffield, UK).

2.1.2 Bacteriological reagents and bacterial strains
Agar and agarose were from Garcia Industries. Tryptone, yeast extract and NZY broth were obtained from Lennox plc (Glasgow, UK). Antibiotics ampicillin (sodium salt) and tetracycline were from Morrison Ltd (Nottingham, UK). Kanamycin was from Cocker plc.

In some research projects you may have made up solutions or other reagents (either standard or specialised) for which you need to give a 'recipe', in which case you can include a 'Reagents' or 'Solutions' section. You can write recipes for solutions in two ways:

(1) either write the volume and concentration of each reagent that is added to make up the solution, and the final volume of the solution:

> Yulug solution:
> 200 ml   5M sodium chloride
> 100 ml   1M sodium citrate
> water to a final volume of 1 l

If you write recipes in this way, remember that if you give volumes, you must also give concentrations; for example, writing '*add 200 ml NaCl*' does not tell us how much NaCl to add, unless you have previously stated that the NaCl is at a concentration of 5 M.

(2) write the final concentration of each reagent in the solution:

> Yulug solution:
> 1 M sodium chloride
> 0.1 M sodium citrate

The five recipes below are written according to method (2), and so, for example, the solution '1 × TBE' is made to a final concentration of 45 mM Tris-borate and 1 mM EDTA. Remember to include the pH of any reagent for which this is important:

2.2 Solutions and media

2.2.1	**General solutions**
1 × TAE:	10 mM Tris, pH 8.0, 1 mM EDTA, pH 8.0
1 × TBE:	45 mM Tris-borate, 1 mM EDTA
20 × SSC:	3 M NaCl, 0.3 M trisodium citrate
Denaturing solution:	1.5 M NaCl, 0.5 mM NaOH
Neutralising solution:	1 M Tris, pH 7.4, 1.5 M NaCl

Writing Conventions

Each discipline has its own way of writing Materials and Methods, and your best sources of information are good recent theses or dissertations in your field, and published papers. However, there are certain general conventions to remember: scientific writing is denser than normal English and avoids many words we would usually use, in particularly prepositions (*of*, *in*, *on*, etc.). So, for example, you would write *5 ml water were added* rather than *5ml of water were added*, or *10 nm diameter*, rather than *10 nm in diameter*.

Take care not to use the colloquialisms of your laboratory or discipline; for example, a chemist would refer to methanol as CH_3OH, using standard chemical nomenclature that is recognisable by everyone, while a biologist might write the colloquialism MeOH which could be meaningless to most other scientists. If you are going to use such colloquialisms, which are non-standard abbreviations, include them in your List of Abbreviations.

Remember to use the correct SI units and abbreviations for times, weights, lengths, chemicals etc. (see *Appendix 7: SI Units and their Multiples* and the List of Abbreviations section in *Chapter 8: Deciding on a Title and Planning and Writing the Other Bits*).

What Tense to Use

You can write your Materials and Methods in either the present or past tense. The present tense can be used for general procedures that are exactly repeatable, the past for specific procedures that are not. In some cases, such as mathematical modelling, for example '2.3.1 the Veronica-Fisher Extrapolation' we gave above, the past or present tense is appropriate. In other cases, for example, the geology method, '2.3.1 Use of Core Samples' the writer would need to use the past tense. Once these core samples have been taken, exactly the same samples cannot be taken again and therefore writing in the present tense, as below, is inappropriate:

2.3.1 Use of core samples
Conventional core samples from well bore 2 (Salisbury Avenue site, depth intervals: 233.87 to 243.23 m, 274.04 to 285.46 m) and well bore 3 (Cecil Road site, depth interval: 222.00 to 229.17 m) are taken. The samples are photographed and logged; records are given in Appendix 2. Samples are taken at intervals of 1 m.

We recommend that you read the section on Scientific Style in *Chapter 14: The Use of English*.

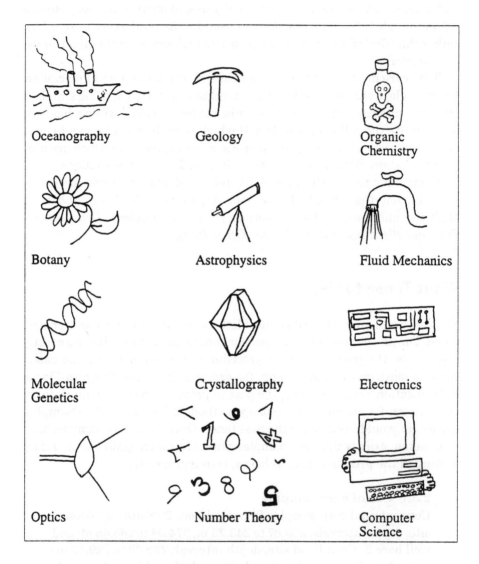

Fig. 3.1   Some materials.

Explaining the Principles

If you feel the need to explain the principles underlying your use of certain materials (or methods) make sure you get it right. In a recent microbiology related thesis a student wrote '*SM buffer is an enriched buffer which allows the growth of phage particles*'. It does not: SM buffer allows the stable elution and dispersal of phage particles, which is something entirely different. This kind of error is a clue to examiners that you are not really familiar with your field and may not fully understand your experimental techniques.

Including the Details

Include all the details of your protocols so another scientist could repeat them exactly, for example, if a solution needs to be of a particular alkalinity or acidity, or if the purity of a chemical is important (some of them are sold at different grades of purity) give the reader the details. You may also want to add useful information about how to make up the reagent/solution/ medium, reminding people to filter sterilise or autoclave solutions, for example:

2.2.2	**Lowell-Badge Medium**
Tomato juice agar:	Add 20.0 g agar to 200 ml freshly squeezed tomato juice. Add 800 ml water and autoclave immediately.

If you are, for example, describing geographical areas, do so precisely, giving map references (where appropriate) and using accepted spellings. If you are dealing with living organisms, give their full taxonomic names, and give details of their sex, strain, age, diet and other environmental conditions, if this is relevant. Also, if appropriate, give the name of the authority or ethics committee that gave permission for the experiments; give similar information if your subjects were people.

Controls

Experimental controls are absolutely essential for checking the validity of data, therefore they must be included in your Methods.

Using Kits

You do not need to copy the instruction manual for techniques in which you have used a kit providing you refer readers back to the manual. However, give a brief summary of how the kit works and refer to any papers containing the methods on which it is based, so the examiners can see you understand what you have been doing—it is essential that you really do understand the principles that the kit is based on. For generally used protocols that occur within the kit write *'according to the manufacturer's instructions'*. You must describe in detail any additional steps or modifications that you have made to the kit. For example:

> Full length cDNA was prepared directly from tissue culture cells based on the method of Cherry and N'Dour (1999) with a SpeediNuck cDNA synthesis kit (Bioshock Inc.) according to manufacturer's instructions. The following modifications were made to the manufacturer's protocol: for the initiation of cDNA strand synthesis, the oligomer used had the sequence $(T)_{20}$; for the insertion of the double stranded cDNA into the vector, the ligation reaction was carried out with T53 ligase (Clonarama Ltd, Aust, UK) at a final concentration of 0.1 U/ml.

For unusual protocols or ones in which the manufacturer's manual has not supplied a reference to a published paper, you might want to put a detailed protocol in an Appendix.

Specialised Equipment

While it is not necessary to describe the use or set up of standard items of equipment, it is a good idea to include a section within your Methods covering any specialised equipment you have used or made. Acknowledge any workshops that helped you and give the specifications of apparatus you made or had made. Describe the reasons for any modelling approach you took, for example, economies of budget may have dictated the use of cardboard or plywood instead of acrylic or steel in the early 'cut and fit' stages. Also reference experimental designs that are based on previous work: if your vortex tube apparatus was inspired by that of Lake and Palmer

(1995) and by the later models of Emerson (1997, 1999), then give them a reference and present your modifications to their designs.

Technical details—for example, calibration results for particular items of equipment—can either be included here or put in an appendix. What you choose to do with technical details depends to a large extent on the conventions of your discipline and how necessary these details are for the reader to understand your experimentation. If the calibration methods are non-standard or particularly error prone you may wish to have a separate section for them as we discuss below in 'Statistical Analysis, Approximation Methods, Artefacts and Repeatablility of Measurements'.

Make full use of diagrams in this section. A picture paints a thousand words and it is often impossible for a reader to fully grasp a design from a written description alone (see *Chapter 7: Figures and Tables*).

Fig. 3.2   Specialised equipment.

Numbers and Symbols

Use symbols in the correct context and correct font style, for example, *C* can denote carbon, part of the symbol for degrees centigrade (°C), the speed of light, and so on. Always use SI (Système International) units and write their correct symbols (we have included a list of them, in *Appendix 7*).

Decimal points

If you do a calculation that results in a value with many figures after the decimal point, round off to a number that reflects the accuracy of your experiment.

Values that are being compared should have the same level of error and accuracy as far as possible: compare those rounded to equal numbers of decimal places, for example, compare 0.005 with 1.000 rather than 0.005 with 1. Be careful with statistical values, for example, if you are giving the mean, and the standard error of the mean, both values should have the same number of digits after the decimal point.

Use decimals, not fractions, wherever possible.

Equations

Make sure the correct symbols are in place, and that you have put everything on the appropriate line in subscript, superscript or normal script. Take care with italics and use them only where necessary, according to the conventions of your discipline.

Calculations

The examiner needs to be able to see that you have understood the problem. Include all the details of any calculations you have made. If you get it wrong, at least the examiner can see why you are wrong and you will get marks for showing your method. Use standard mathematical notation and put in enough explanation so that the reader can follow what you are doing easily. Begin each new stage of your calculation on a new line and use linking words like *and, so, therefore,* to guide the reader through your thinking. Make it absolutely clear what units you are using (if any) and what a number or set of numbers represents: if 5 is the number of petals that are red, then state this explicitly, if your units are Ω, make this clear. It is a good idea to check the results of your calculations are the right order of magnitude and in the correct units, for example, m^2 rather than m^3. It is easy to make mistakes with calculators and get decimal points in the wrong place. If you have a lot of data it is worth using a graph or table to provide a simple display for the reader.

Statistical Analysis, Approximation Methods, Artefacts and Repeatability of Measurements

Errors exist in any experimental system. Statistical analysis is an objective method for avoiding, or detecting, biases in data. Always try to include a statistical analysis of your data, estimates of errors, and approximation methods. Discuss the reasons why you chose particular methods, and take care to carry out the appropriate calculations. Use the correct words and symbols, for example 'standard deviation' and 'standard error of the mean' are two quite different values. Ask your supervisor for advice.

For some projects you may also need to include a section on artefacts. Artefacts are results that arise as a by-product of the methodology you have used, rather than useable experimental data. A discussion of artefacts, or other likely sources of error is particularly useful for projects in which you have developed new techniques. Some experimental approaches may result in a skewing of the data, and you should think carefully if this applies to any of your work.

List any factors that might make it difficult for someone to repeat your experiment. This will be more of a problem with some experiments than others, for example, geological field studies may have many more problems of repeatability than a laboratory based chemistry or physics project.

Computer Programs and the World Wide Web

If you have written computer programs, these may fit best in an appendix. Reference any available computer programs that you have used, including their version number, by referring either to published papers or by giving the manufacturer. Give the name and addresses (http://www...) of any sites on the World Wide Web that have provided you with Materials and Methods.

Hazards

Many commonly used reagents are dangerous so include all appropriate warnings.

Questionnaires and Other Relevant Information

If you have used questionnaires to collect data, include a copy of the questionnaire, and have this translated into English if the original is in a foreign language.

Acknowledging Other People If They Helped You

If other people provided significant help for certain experiments, carried out part of an experiment, or provided samples that you worked on, you must acknowledge their assistance. Include your helper's name and their university or college if it is different from yours. You could write '... *ethanol extraction experiments were performed with the aid of Professor C. Phillips*', or '... *samples were provided by Dr B. Lo, Merton University*', or '... *resonance data for the Every analysis were kindly provided by Dr P. Hopkins*' .

Figures, Tables and Appendices

By 'figures' we mean illustrations that are included in the text, for example, line drawings or photographs of a piece of equipment, or flow charts of processes. It is often far easier to get the point across by using a figure rather than relying solely on a written explanation, so use figures wherever they will make things clearer. Take care when writing on the data from your experiments, such as X-ray films, traces, photographs, etc. which you might use later as figures in your thesis or dissertation. If you write notes on your data, do so neatly, so that you do not obscure any important details and so that the writing can be edited out if you need to use these data as a figure.

If you are using a particular notation to create your figures, such as chemists use to describe molecular configurations or electron orbitals, bear in mind that these are usually not standard, so explain your notation fully.

Tables are the most efficient way of presenting a large bulk of information, for example calibration readings from instruments, so make use of them.

Some departments favour placing the Materials and simple, generally used Methods, such as computer programs, recipes for buffer solutions, proofs, into an Appendix. See a good recent thesis for guidance on your departmental conventions, and see *Chapter 7: Figures and Tables.*

Reading What You Have Written

When you have written your Materials and Methods, try imagining you are a scientist in another country who knew nothing about the experiment, but had to repeat it. Could you do this from the instructions you have written? Get a friend who knows something about your field to read this section, and see if they can follow it.

Common Mistakes and Points to Bear in Mind for Different Disciplines

- Do not use different non-standard abbreviations in different places for the same materials or units, etc.
- Give details of the computer programs or databases that have been used.
- Use the correct Greek letters such as α or β, rather than incorrect Latin letters such as a or b
  An unfortunate student who had written ul for μl in their thesis was made by an examiner to replace every single ul in the thesis (several hundred in all) to μl. Beware
- When giving the composition of reagents and referring to percentages, for example '20% ether-petrol' or '0.6% acrylamide', be clear whether you are referring to volume:volume or weight:volume percentages
- Do not capitalise the names of chemicals; for example, Sodium Chloride and Sodium chloride are incorrect, sodium chloride is correct
- When writing formulae be careful *always* to use the appropriate subscripts and superscripts, for example:

$$C_{44}H_{62}O_6SSi_2 \text{ not } C44H62O6SSi2$$

- When writing chemical names be careful to italicise appropriately, for example

  1-(*tert*-butyldimethylsilyloxy)-1-methyl-3-(phenylsulfonyl)-4-butanol

- Do not use colloquialisms because your meaning may be mistaken, for example, never write 'hot' when you mean 'radioactive'. Do not use 'rpm' instead of 'xg' when describing centrifugation protocols. This is because every centrifuge is different and the force on the sample depends on revolutions per minute (rpm) and the radius of the centrifuge. The x gravity (xg) is the measure of centrifugal force, and is commonly understood by everyone. The only time it is ever correct to refer to 'rpm' is when you also give the exact make and model number of the centrifuge and the rotor, so that people can replicate your conditions precisely

- When describing centrifugation protocols, use xg instead of g; 'xg' is times the constant for gravity and refers to the force on your sample, 'g' is grams
- When you are trying to describe actions, use the correct word. For example, do not use the word 'spin' if you mean 'centrifuge'. We recently encountered a sentence that included 'when the sequence was fed into the database ...'. Use correct English, and 'enter sequences' rather than 'feed' them. Another common mistake is to 'run' samples rather than 'electrophorese' samples. This may seem trivial, but again, one of us was examining a thesis and came across the phrase 'the tube was pulsed'. What does this mean? It transpired (in the examination) it meant that the tube was briefly centrifuged at high speed
- Learn about the conventions of notation in your field. For example, human gene names should be written in italicised capital letters (*CTBP2*, *VCP*) whereas mouse gene names should be written in italicised letters, only the first one of which is capitalised (*Ctbp2*, *Vcp*)
- When writing species names of organisms, convention dictates that these are written in italics, with a capitalised genus name and a species name that begins with a small letter; for example, *Homo sapiens*, *Mus musculus*, *Rattus rattus*. You can, after the first mention, abbreviate this to, for example, *H. sapiens*, *M. musculus*, *R. rattus*

Key Points

- Write a simple and technical account of what you did
- Do not include any results
- Always use SI units, with the correct abbreviations, without full stops; see the table of SI units in *Appendix 7*
- Non-standard abbreviations such as 'RT' (room temperature) must be listed in an Abbreviations section at the front of your thesis or dissertation

- Be consistent about spacing between numbers and units, for example,

  45mM or 45 mM

  either is fine, as long as you are consistent
- List exactly the correct materials and methods including, for example, any important details such as pH
- State exactly which piece of equipment you used, because this may also affect your results; give the name, model number, and manufacturer
- Reference computer programs, databases and World Wide Web sites
- Make use of figures and tables, they are the most efficient way to convey bulky detailed information

Chapter 4

PLANNING AND WRITING THE RESULTS

In your Results you are giving a brief outline of each experimental strategy (not a detailed protocol), then telling the reader exactly what happened and what you learnt. You are not giving detailed procedures, which are in Materials and Methods.

Your presentation of results will depend on the nature of your research. You might put your results in one or more separate chapters, or, alternatively, as a section within a chapter covering one aspect of your research. However they are presented and whatever they are called, your results are the core of your thesis. They show the examiner what you did and how you did it.

Not all theses or dissertations, particularly in theoretical subjects, have a Results section. If existing theories or findings are being reviewed it can be more sensible to integrate them into the Discussion chapter. Have a look at a good recent thesis or dissertation in your field and talk to your supervisor for further advice.

Literature Survey

Although you will, or should, have a set of references collected during the course of the project, before starting on your Results it is a good idea to carry out a literature survey: a review of all the literature (books and journal articles, for example) that might be relevant to your project. In more theoretical subjects it may be impossible to write the Results chapter without doing a literature survey, because the materials you are working with are theories and equations and so on, from published papers and circulating pre-prints. In practical subjects you could leave the literature survey until you write your Introduction, but doing it now has a number of advantages: reading the literature will give you an idea for what to include in the Results, and appropriate styles of writing, it will also bring up ideas that you had not thought of, and may help you to understand the importance of your work and the area in which you have been researching (see *Chapter 12: Resources*).

Read all your references and note the key facts so you have them easily to hand for writing your Results (and other chapters). Remember that any references you cite should be entered into your computer reference file or database program (see *Chapter 2: References*).

If you are using the World Wide Web as a source of information, reference the Web pages by putting in the address so the reader can see where your information came from.

Planning Your Results Chapter

Unless you were incredibly meticulous in the initial planning of your project, and everything went exactly according to plan, the order in which you did the experiments will not be the order in which you present your results. Your results need to be logically rather than chronologically arranged, so that both you and your reader can easily grasp their significance and relationship to each other. You need to do two things to plan your Results:

* clarify your aims
* arrange your results to support your aims

The whole point of taking time over your planning is to get things right before writing the text in full.

Clarify your aims

To plan your Results chapter you need a clear idea as to the aim of your project. Your results need to be organised so they support this aim. Whether writing an undergraduate dissertation or a PhD thesis, your aim has probably changed slightly during the course of your research in response to either your own or other people's work; you might not have quite the results you were expecting, possibly someone beat you to it and you had to rethink your project in the light of their findings, or you could have stumbled across a much more interesting avenue of research along the way. Whatever your situation, your results are your results. Remember that your aim should fit your results rather than the other way round. If your results do not seem to bear much relationship to your original aim now is the time to rethink your aim.

Worst Case Scenario … a bunch of meaningless results. The best thing to do is to accept the situation and then look at why your results are meaningless. Possibly if you compare your research with established work in your field you will be able to explain why things went wrong. It may have been because of badly designed experiments or perhaps it was just bad luck—science is like that sometimes. If you provide a critique of your methods and carefully analyse and explain your results you should be able to produce a reasonable dissertation or thesis. You may find you need a total overhaul of your ideas, in which case it is best to scrub your original aim from your mind and start again. It is a lot easier to do this with someone else, so go through your ideas with your supervisor.

Arrange your results to support your aims

To plan the order in which to present your results you need to know what results you have. Think about your project and what you have achieved. Read through your practical books and any other notes that are relevant to remind you of your experiments. Note the aim and result of each experiment in a file on your computer or on paper. Once you have made these notes you can arrange your results into a logical order.

What Results to Include

Include all results that are relevant to your aim: the major results (around which you build each chapter or section), and the minor results, such as making or testing reagents, or calibration results, that support the major results. Remember to include all data from controls as these show the validity of your experiments.

Do not exclude results that contradict, or might contradict, your aim. (Honesty is not only good for the soul, it is also good for people's careers.) If you have contradictory results, attempt to explain them. The same is true for artefactual results produced as a by-product of the methodology. Contradictory and artefactual results arise in all experimental systems. If one of your experiments did not work and you have managed to discover why, it is worth going into a detailed explanation, so that the examiners can follow your reasoning and other scientists can learn from your findings.

What Not to Include

Do not include results that are or have become irrelevant to your aims. If you do include them you will not only waste a lot of ink and paper, you will also waste time—yours and your examiners', who will wonder if you really understand the significance of your own work.

Just as you should never include results in your Materials and Methods, you should not include materials and methods in your Results. Do not present any detailed protocols, just a brief overview of your approach to each experiment.

Including Other People's Data

This is fine as long as you reference it clearly. By doing so you give credit where it is due and avoid any charges of plagiarism (and potentially failing your degree). Cite published or unpublished data, and if someone helped you with a technique or provided you with a reagent give them a reference as well.

The Order in Which to Present Your Results

To arrange your Results chapter you need to decide on your main results, then decide on the order in which to present them. Start with the results that are the simplest and underpin your other work. Then, once you have set these down and are on solid ground, move to the next result, making sure it is supported by your previous work, and repeat the process. 'Walk' from result to result like this, setting out a logical pathway for your reader to follow.

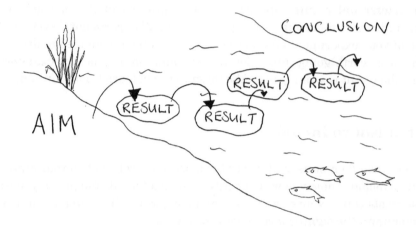

Fig. 4.1   Walk from result to result, setting out a logical pathway for your reader to follow.

Arranging your supporting results into groups under your main results

Your supporting results are those from the many small experiments that you did in order to get the information you needed to do your main experiments. Present these results starting with the simplest followed by those that build on the data you have generated. Group each of these small results along with the main experiment they supported.

How Many Results Chapters, in Which Order?

Once you have ordered your results into coherent groupings under each main result it is worth considering how many Results chapters to write. We have never come across any institution that has a rule about the number of Results chapters you are allowed. If you have a number of markedly different main results, form a separate chapter around each of them. For example, in a computing thesis, you might have two chapters, one describing the program you have written, the second putting that program into action. In molecular biology, a project with tissue culture results, genetic mapping results, and developmental analysis, will divide sensibly into three chapters. A project that has used standard molecular genetic techniques throughout would work well as one chapter. Generally, the larger your thesis the more sensible it is to have a number of Results chapters simply because it is easier to digest the information if it comes in small well-defined chunks. If you have more than one Results chapter, order them so that related topics are next to each other.

Writing Your Results Chapter

When you have planned your dissertation or thesis it will be time to write your Results chapter. Your plan is there as a blueprint for your writing. You do not have to stick to it absolutely but use it as a structure that you can change if necessary.

In your Results you are giving facts not opinions. You are telling the reader which experiments you did and what happened. The prose of the Results chapter should flow smoothly from statement to statement so it builds rationally to support your aim. Make use of headings, figures and tables to break up the data into easily readable sections so that both you and the reader can keep track of your results.

When writing your Results state which method you used, and refer to the specific section of Materials and Methods which describes this protocol. It will be helpful to the reader if you have arranged your materials and methods into the same order as your results—as much as is possible.

It is probably best to write mainly in the past tense. If you use the present tense there is a danger of appearing to make unfounded generalisations, for example:

> ... a set of 36 related esters were chosen and analysed according to the Rantzen method, 5 of them were found to be positive, 29 were negative and no results were obtained for 2 of the samples...

This tells the reader your results—what you did and what happened. If we change this into the present tense:

> ... a set of 36 related esters are chosen and analysed according to the Rantzen method, 5 of them are found to be positive, 29 are negative and no results are obtained for 2 of the samples...

This sounds as if you are making a generalisation that for this particular experiment, 36 esters are always chosen and analysed by the Rantzen method, 5 are always positive, 29 are always negative and 2 never give results.

When presenting data, make full use of all the information that comes out of each result and explain its significance to the reader. For example, in a study of electrolytic corrosion, if you find that the mean average weight of two sets of alloy ingots has dropped, you could simply say,

> 'The final mean average weight dropped from 55.0 g to 53.2 g for alloy A and to 46.5 g for alloy B.'

but it is much easier for the reader to follow if you also give an interpretation

> 'The final mean average weight for alloy A was 53.2 g, which was significantly higher than that of 46.5 g for alloy B.'

Here we have not just presented the raw data to the reader, we have also worked with the data to make a useful comparison.

Writing Style

Use accepted scientific terms, and avoid jargon and scientific colloquialisms used around the laboratory. Use the correct units and their abbreviations. If you are struggling with the writing, try looking at published papers and talk to your supervisor (see *Chapter 14: The Use of English*).

'Strategy' Section and 'Summary' Section

In some disciplines it is common to start the Results chapter(s) with a section explaining the experimental strategy used. This can be useful for setting the scene for your Results chapter(s) because it tells the reader the main results in advance, and makes it easier for them to grasp their significance. If you write a 'Strategy' section, keep it short and to the point. The reader does not want a detailed treatise on the philosophy behind the approach. Do not muddle the strategy with the results, which follow. Alternatively, a bullet point summary of findings at the end of each Results chapter is used in some disciplines.

Titles for Results Chapters

You could simply call your chapter or section 'Chapter 3: Results'. However, a brief self-explanatory title would be more useful, especially if you have more than one Results chapter; you might want to write something like, 'Chapter 4 Results: Assessing the Bond method of processing liquid-sodium coolant' or 'Chapter 4 Results: Vaporisation of zinc alloys'. You do not have to use the word *Results* in your title and could have a heading like 'Chapter 3: Mass spectroscopy of Palaeolithic bone samples', or, 'Chapter 5: Maser derived observations of electro-magnetic emissions from Betelgeuse'. Exactly how you title this chapter will depend on personal taste and the conventions of your discipline.

Preparing for Your Introduction and Discussion Chapters

You will probably find that as you write your Results chapter you think of points that should go into your Introduction and Discussion chapters. Keep a note of these as you go along, either on a computer file or in an Ideas notebook for future use.

Statistics and Numbers

You may have to carry out detailed statistical or other numerical analysis of your data in the Results chapter. We are not going to provide rules and guidelines for the hundreds of diverse types of number crunching that readers of this book will have to undertake. Your supervisor is the best person to comment on your individual needs. But, there are a few general points to make about presenting numbers and calculations in any dissertation or thesis, which are discussed in the 'Numbers and Symbols' section of *Chapter 3: Materials and Methods.*

Figures, Tables and Appendices

Think carefully about which data to place in figures, tables and appendices; these can all be very helpful ways of summarising and showing your results. Particularly in practical subjects, most of your results are in the form of data that could be presented as figures or in tables. Your text is there to introduce and explain the figures and tables and to describe data they contain. Do not reiterate the contents of the figure or table in your text, simply discuss the important points and refer the reader to the appropriate table or figure for more information. For layout of figures and tables, see *Chapter 7: Figures and Tables* and for appendices see *Chapter 8: Deciding on a Title and Planning and Writing the Other Bits.*

Common Mistakes

- Including materials and methods
- Including duplicate or irrelevant results
- Writing results in a chronological not logical order
- Sticking with original aims, when the project has outgrown them

Key Points

- Carry out a literature survey before writing your Results
- Plan your Results by:
    clarifying your aims
    arranging your results to support your aims
- Arrange your results logically and move from one result to the next related result
- Write multiple Results chapters arranged around different subject areas of your project

- Make use of tables, figures and appendices to summarise your data
- Keep your style crisp and to the point, give facts not opinions
- A bullet point summary of each Results chapter can be helpful

Chapter 5

PLANNING AND WRITING THE INTRODUCTION

Structure of the Introduction Chapter

Your thesis has to do more than simply supply the reader with data and a hint or two as to how you came to your conclusions. You have to take the reader firmly by the hand and guide them step by step through your experimentation and the reasoning that brought you to your conclusion. The first step in that process is introducing them to your project, which is what your Introduction does.

In your Introduction you have to lay out the background of your research and show the relationship between your work and the wider field. You also have to tell the reader why your project is interesting and what question you have aimed to answer. You need to present your broad area of research and then narrow down to your specific interest, and finally to pose the question that your thesis is answering. You could visualise your Introduction as a funnel.

Whether you are writing just one introduction, or a series of smaller introductory sections to different results chapters, the principles are the same: start broad, end narrow.

Literature Survey

Before you start planning your Introduction, carry out a literature survey, if you have not already done so while planning your Results (see

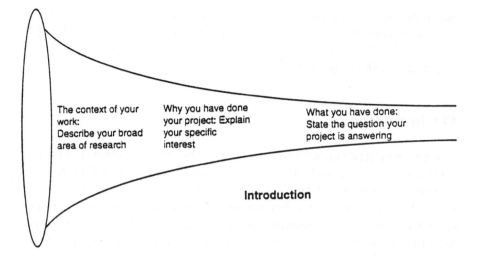

The context of your work: Describe your broad area of research

Why you have done your project: Explain your specific interest

What you have done: State the question your project is answering

Introduction

Fig. 5.1　You could visualise your Introduction as a funnel that draws in the reader.

Chapter 12: Resources). Look through any literature that might be relevant to your project so you do not miss any important papers. Track down any references you know you need, for example, the original reference for a method that you used, or the first paper announcing an important finding in your field.

When you have finished your literature survey read through all your references; this will help to give you a thorough understanding of your research and usually throws up new ideas about the significance of your Results. Note relevant points for your Introduction as you read through the references.

Planning the Introduction

You need to have your aims clear in your mind before building an introduction to them. Note the key ideas you had while writing your results plan and the main points from your literature survey. Use these notes to plan your Introduction. You will probably not get your plan right the first time around so your key points need to be in a form that is easy to arrange and rearrange. You could sketch your plan on paper, write each point on index cards and then shuffle them around until you get them in the right order, or shuffle the points around on your word processor. At this stage, only pick out the major points.

The beginning

In your opening paragraphs you need to give the reader a precise but general overview of the area in which you are working. Cover any important findings or theories which led to your project or which affected your work, remembering to note any references you need. Bear in mind that a reader cannot ask you for explanations as they read through your text, so you need to be as explicit as possible and provide all the information you think a scientist in your field might need to understand your project.

There are two common ways of structuring the beginning of the Introduction. You could provide an account of your field at the time of writing (reviewing the current status of findings and theories, the *status quo*), or give a brief history of your field up to the time of writing (including the current status of findings and theories). If we take the example of a thesis written by a gardener who has been planting flowers as part of a landscaping project. The gardener has developed what they believe is the best combination of flowers and scents for a large garden. Using the first approach, their thesis would begin by saying what a garden is, then stating the architectural principles behind landscaping, and finally describing various types of flowers that could be planted, perennial, biennial, annual and so on. Using the second approach they would start by giving an historical introduction to gardens—the introduction of different types of flowers through the ages, and the first attempts at landscaping followed by the development of certain landscaping principles. Use whichever approach you prefer, *status quo* or 'historical', but the former, giving an outline of your specific area of study as it stands at the time of writing, is probably easier.

Remember you are introducing the reader to the small part of your science that you have been working in, you are not writing a textbook about your discipline or a history of Western Science.

Literature survey or background chapter

In some disciplines the Introduction is split into two chapters, one called 'Introduction', which deals specifically with the details of the project, and one reviewing the literature about the area of interest ('Literature Survey' or 'Background'). You can write the Literature Survey in the same way as we have outlined above, either a current day account or an historical introduction—see a good recent thesis or dissertation in your discipline and talk to your supervisor for guidance.

The middle

In the middle section of your Introduction you have to take the reader from the general to the specific—your aims. You have to tell them what particular aspect of your field of research your project investigated and why. Give the reader an up-to-date outline of any findings and theories that are relevant to your project. Again, remember that a reader cannot ask you questions while they read, so you need to include all the information they will need to understand your project.

Strictly, every statement of fact you make should be referenced. In practice this is impossible, but you should reference the key points and any ideas that are not either widely known or commonly accepted, so your examiners can check your sources. It is easy to under-reference, and difficult to over-reference a thesis. If you think you are in danger of either, get advice from your supervisor. You have to demonstrate constructive criticism of the references you are using, comparing and contrasting different theories and findings where appropriate. Remember to reference Web sites (by giving the name and address) as well as published papers.

If it is possible, or relevant, the structure of your Introduction should reflect that of your Results chapter. Introduce the various aspects of your research in the order in which they are presented in your Results—using the plan of your Results as guide.

If your research concerns a new method, piece of apparatus, or computational method, then explain the problems with the old methods and apparatus. When you come to describe your improvements in your Results and Discussion, the reader will understand why an improvement was needed and why your approach is better.

Dealing with awkward points and comparisons

If you are working in an area where there are a number of competing theories, processes, or findings, you will need to show the examiners that you are aware of all of them. If you put forward a theory which is perhaps quite controversial, show you are aware of the counter-arguments and alternative interpretations of data. When comparing complex theories, processes, or findings it is best to go through the comparisons point by point. It is a lot easier to grasp the comparison if you run them side by side rather than setting out one theory in full followed by the other.

The end

The end of your Introduction is the simplest part to plan. Here you just need to tell the reader your aims—the question you are attempting to answer:

> The aim of this project was to determine if the increased incidence of dysfunctional reproductive behaviour among landfish in the Milton Keynes area is caused by stray radiation from domestic television sets.

You could also give a brief introduction to your Results chapter: outline your experimental approach. A few simple sentences are all that is required. If you have more than one Results chapter, note what is in each one so that the reader knows what to expect. You could put these notes in bullet points, which will look neat and be easy to understand at a glance.

Figures, Tables, and Appendices

Use figures wherever they will clarify your argument. In the Introduction you may wish to use figures from published papers or books; this is perfectly acceptable providing you fully reference them. Tables and Appendices are the most efficient way of presenting a bulk of detailed information. Make use of them; again, you can use information from publications providing you reference them fully. We discuss figures and tables in *Chapter 7: Figures and Tables*, and appendices in *Chapter 8: Deciding on a Title, and Planning and Writing the Other Bits*.

The Importance of Planning

Do not hurry the planning of your Introduction. It sets the scene for the Results and Discussion chapters. A well laid out Introduction, along with well planned Results chapters, will make the Discussion a lot easier to plan.

Writing Your Introduction

In your Introduction you not only have to make sure what you write is relevant, honest, rational and can be backed up by evidence, you also have to sustain the reader's interest through your argument. Your Introduction should give the reader a sense of setting off on a journey of discovery. You are taking them by the hand and leading them into the world of your research, they might have been there before, or they might not—even if you are taking them over fairly well-known territory, there is no reason the trip should be boring. You are interested in your subject and the background to it. There are lots of fascinating findings and theories, some of them perhaps quite controversial. To keep your reader's attention your prose needs a feeling of dynamism. Keep your sentences quite short; this will keep the energy up. Each sentence should follow logically from the previous one and towards the next. Similarly with your paragraphs, do not make them too long, and stick to one main idea in each one.

Try to avoid wordiness, and keep clear distinctions between statements of fact and statements of opinion. Include only relevant information. If you need to, define your key terms at the beginning of your Introduction, and understand specialist terminology, because its misuse will make the examiner question how well you understand your subject and look out for more serious flaws in your thesis or dissertation. For further advice see *Chapter 14: The Use of English*.

Plagiarism

Never copy anything from other people's papers or work without fully acknowledging and referencing it. Some writers think they can get away with it here and there and cover themselves with other people's glory, but examiners know the literature as well as anyone and are likely to recognise plagiarised work—it could even be theirs. It is obvious when someone's writing style changes, which puts the examiner on the alert for plagiarism. Students are rightly failed for plagiarism.

Fig. 5.2   A writer found guilty of plagiarism.

Common Mistakes

- Putting in too little information for the reader to understand the rest of the thesis or dissertation
- Including too few references to support the statements that are being made
- Scattering information around the Introduction in no logical order, so that the ideas do not flow from paragraph to paragraph, section to section

Key Points

- Carry out a thorough and up-to-date review of all literature covering your area of research
- The beginning: start broad and set the scene
- The middle: narrow down to your chosen speciality, and set your work in context
- The end: ask one question that is addressed by your Results, this is your aim
- Keep your writing crisp, to the point and dynamic

Chapter 6

PLANNING AND WRITING THE DISCUSSION

Structure of the Discussion Chapter or Section

In your Introduction you told the reader what you were going to do and explained the background to your research. In your Results you showed them what you had done and how you did it. Now in your Discussion you have to tell them what it all means: the relevance of your research and the conclusions that can be drawn from it. Also outline future work that could be carried out. Whether you have a single Discussion chapter, or Discussion sections at the end of each of your Results chapters, the Discussion section really should be a discussion of your project, and not simply a list of conclusions.

In your Introduction you started broadly and narrowed down to your aims. In your Discussion you start narrowly and broaden out. Begin by restating your aims, then consider your methodology and results, and go on to build a wider picture of how your results fit into the context of your field of research. You could visualise your Discussion as a funnel.

Read the Plan of Your Introduction and Results, and Your References

Your Introduction and Results are the building blocks of your Discussion, so take time to read through your plans of them, and relate your work to that discussed in the references from your Introduction.

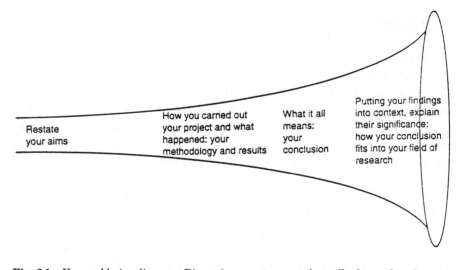

Fig. 6.1 You could visualise your Discussion as a trumpet that tells the reader what you have achieved.

Start by considering your results. Think about the details of your experimentation, why you chose particular methods, whether the experiments or calculations could be improved, and what each one told you. Then consider your results in relation to the wider field of your research—the information you have gained from your references and given in your Introduction. Take notes of all the points you want to include in the plan of your Discussion.

Planning the Discussion

The beginning

Your opening paragraph should start by introducing your aims again:

> The aim of this project was to determine if the increased incidence of dysfunctional reproductive behaviour among landfish in the Milton Keynes area is caused by stray radiation from domestic television sets.

The middle

Now add points for discussion of individual results to build up a body of information that addresses your aim. Remind the reader why you carried out each experiment or calculation, and why you took certain approaches; present minor conclusions from each experiment or calculation as you go along. You may also want to tell the reader how you could have improved the experimental aspects of your work, or your approach to theoretical problems. This is especially relevant if your project has not been particularly successful or a new technique has been developed that would have been useful. Next relate your results to your field of research, citing the specific references which cover the area you are considering.

If possible your results should be discussed in the order in which they were presented in your Results section(s) to make your whole argument easier to follow. Make speculations if you wish, but only if they are soundly based in your research and are clearly distinguished as speculation rather than statements of fact.

Dealing with awkward points and comparisons

There may be, particularly with theoretical research, counter-arguments and alternative interpretations. Be aware of these and deal with them immediately they become relevant to your discussion; alternative views will occur to your examiners immediately. If you leave these points to ferment in your examiners' minds they will be distracted from your argument, wondering if you have thought of the other possibilities. Acknowledge the problem:

> While there is some dispute about the existence of landfish in the M4 corridor...

and then deal with it straight away:

> I have found of landfish spoors which point conclusively to their presence in the area around Slough.

The end

Different disciplines and departments organise the end of the Discussion in different ways; your best guide is a good recent thesis or dissertation in your field from your department. But, the same information goes at the end of almost all Discussions, however they are organised: there is a brief statement as to whether or not the original aim has been addressed, followed by suggestions for future work so the dissertation or thesis ends on a positive note. In some disciplines this information is included at the end of the Discussion, in other disciplines it is put into a separate section called something like 'Concluding remarks and future work'. Alternatively, the Discussion finishes with the ideas for future work and is then followed by a short final statement called Conclusion, which gives a brief summary of the whole project and how it addressed the original aim.

The conclusion

You will almost certainly have been working with an idea of the conclusion throughout your project. As you planned your Results and Introduction it will have become clearer. Now that you are planning your Discussion, your conclusion should be crystal clear. Make sure your conclusion can be backed up by your data; it is best to do this before possibly wasting time writing a Discussion that does not in fact lead to your conclusion. Plan, and look carefully at the results you have; they may be telling you things you have not thought about.

If you have answered the original question posed in your Introduction then explicitly tell the reader so. If you were unable to answer the question then say why. If your work has specifically addressed any of the points from your references, make it clear what new light your conclusion throws on these established findings or theories—note down the main thrust of your argument at each point, making sure you include anything that either supports or contradicts your argument and making clear why it does so.

For some research projects, particularly those carried out over only a short period of time, you may have no useful results. If this is the case for you, your conclusion will be that you could not address the aim of your project. This is perfectly valid providing you discuss intelligently why you were unable to address your aims. This situation usually occurs because of problems in the experimentation, either due to the experimental approach and design, or because of bad luck—the experiments just did not work, even though they should have done.

Suggestions for future work

Once you have settled on a conclusion, jot down ideas for suggested future work and how you might approach it. Show that you can recognise the important questions arising from your work and have a realistic understanding of the ways to address these questions.

Suggestions for future work are particularly helpful in upping your chances if you had a disappointing project. If you were unlucky with your results, or it was only towards the end of your project that you realised you were slightly off target with your research, or the project was ill-conceived and you did not realise this until it was too late to do anything about it, your suggestions for future work will show the examiners that at least you have learnt from your misfortunes or mistakes (which happen to everyone) and can see more productive avenues to follow.

The Importance of Planning

It really is a good idea to plan your Discussion thoroughly before writing it in full, otherwise you can tie yourself in knots. It is very painful to delete words and paragraphs you have laboured over deep into the night, however much they need to be discarded, and it is very easy to become mesmerised by your own words, especially if you are beginning to panic and are running short of time. When something is written down in full, particularly if it is printed in a nice font and well laid out, there is a tendency to believe it more than if it were sketched in note form. Examiners will not feel so indulgent.

You might find you want to slightly rearrange the plan of your Results sections and Introduction while working on the plan of your Discussion chapter or section. If so, do it. The whole point of planning is to get a blueprint that works. Be ready to go back and re-jig your aims and Introduction if necessary, so that your conclusions grow rationally from your original question and the results of your experimentation or calculations.

Writing Your Discussion

You will write your Materials and Methods, and probably also your Results in a fairly terse and minimalist style. Your Discussion, like your Introduction, gives you a chance to write in a more descriptive and fluid style, explaining both what you have done and why you did it. Your prose should be easy to read, so lubricate it with linking words such as *and, therefore, but, however,* to join your ideas together. 'Signpost' your writing so the reader can easily see where it is going. Use headings for the main points of your presentation, and refer the reader to figures and tables in the text. Use the present tense for any general conclusions and the past tense to talk about your results. As much as possible try and write in the active because this is more direct and avoids wordiness:

> I carried out the experiment...
> The ions passed through the cell wall...

rather than the passive:

> The experiment was carried out...
> The cell wall was passed through by the ions...

Some people think that the 'scientific style' requires the passive; this is not so. Having said this, check to see what the conventions are in your field, as some examiners might feel more comfortable with a passive construction (see *Chapter 14: The Use of English*).

Figures, Tables, and Appendices

Use figures and tables wherever they will help the reader understand your presentation. They are particularly helpful if you have a large number of comparisons to make, or wish to illustrate a point. It is unlikely you will have much need for appendices in your Discussion, but use them if you need to. See *Chapter 7: Figures and Tables*, and *Chapter 8: Deciding on a Title, and Planning and Writing the Other Bits.*

Common Mistakes

- Insufficient discussion of the theoretical and experimental approaches taken
- Lack of organisation and insufficient flow from specific experiments or calculations through to the broader picture
- Not addressing key points raised in the Introduction

Key Points

- Start your discussion by outlining the general thrust of your argument: restate your aims
- Then review your results
- Be aware of awkward points and deal with them immediately
- Finish with your conclusion and suggestions for future work

Chapter 7

FIGURES AND TABLES

We use figures to illustrate and clarify a piece of text (for example, a photograph of a plant) and to display data in a pictorial or graphical format (for example, a graph showing rates of vegetative propagation in plants). We use tables to present large amounts of precise numerical and qualitative data.

Figures and tables need to be clearly organised. Make sure you know what information you are trying to convey, and plan your figure or table so that you do so efficiently without including redundant ideas or data. Do not try to cram in as much information as you can into one figure or table, as this will make it difficult to follow. Do not repeat information from your figures and tables in the text, simply refer to the point you want to highlight. Remember to make clear where the data that have gone into a figure or table came from.

Size

The simpler the figure or table, the smaller you can make it; but it has to be easily readable and look good on the page, so do not make it too small. With a more complex figure or table you will need to make it large enough to clearly present the information. Try not to make your figures or tables larger than an A4 page or you will have to fold them over to fit in your manuscript; if you really do need to have such a large figure, then do so.

You can put more than one figure or table on a page, but leave a large enough margin around each figure to clearly separate them.

Numbering

The easiest way to number your figures and tables (both for you and the reader) is by giving the chapter number and then an individual figure or table number; for example, in Chapter 2, the figures (including any graphs) would be numbered Figure 2.1, Figure 2.2, Figure 2.3 and so on; the tables would be numbered Table 2.1, Table 2.2, Table 2.3 and so on. If there are two or more parts in a figure, number and describe them separately in the figure legend, for example, Fig. 2.4(a), Fig. 2.4(b), etc.

Other People's Figures and Tables

It is fine to use other people's figures and tables, and modified versions of them, providing you reference properly, either in the figure or table legend, or on the figure or table itself. If you have modified someone else's figure or table, make this clear.

> Fig. 1.5 is taken from Orme and Wilson, 1999
> Fig. 3.6 is based on a figure by I. Cutler (Stanshall *et al.*, 1996).

If you are using published work, such as maps, check to see that you will not be infringing any copyright (see our section on copyright in *Chapter 12: Resources*).

Figures

Figures, including graphs, need to be as carefully planned as your text. Put in figures that help clarify any ideas or information you are presenting. These might be diagrams summarising our current state of knowledge of the system within which you are researching, they could be flow diagrams describing the strategy you took, they could be photographs of apparatus or geological features or insects' wings—whatever helps explain your text. Figures are, of course, also essential for showing your primary data to your examiners. You do not need to show every piece of data you have produced during the course of your project, but you do need to let the examiners see your important results, those that enabled you to draw a conclusion, and perhaps go on to the next experiment. You need to show experimental data that you are discussing in the text. If you have carried out a set of similar experiments and have a lot of examples of the same kind of data, choose a representative set of pictures to illustrate what you have done.

Before you start to make your figures, read your thesis plan and make a list of what figures should be included.

Creating Figures

Figures are created in a number of ways depending on the type of data, the figure you wish to produce, and the computing facilities that are available to you. Before you start to create a figure, draw a rough draft of the contents and layout. This will help focus your ideas and give you an idea of the relative size of the components and the complexity of the figure. Making illustrations can be very time-consuming so you do not want to waste time revising them. Although all figures must be referred to in the text, they should not depend on the text to be understandable. When planning figures there are four considerations to take into account:

1. The graphical display (does it look good and show what it ought to)
2. Annotation to explain features in the figure
3. The figure title
4. The figure legend (a short piece of text that tells the reader what they are looking at)

Generally it is unnecessary and time wasting to draw a border around each figure, and few scientific journals do this.

Figure 7.1 shows the same data treated in three different ways. One is good, one is bad, one is ugly and bad. Figure 7.1(c) is complete because it includes the above four features.

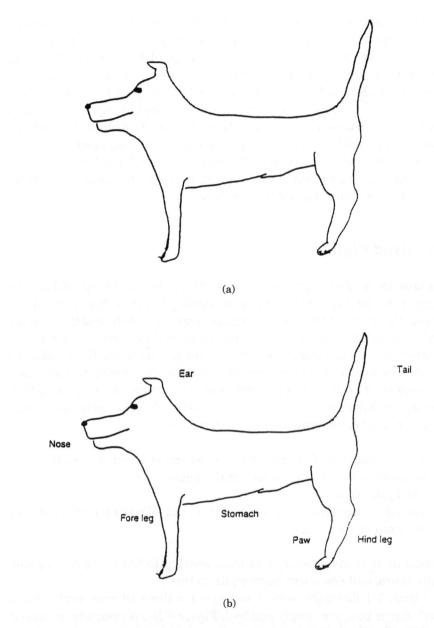

(a)

(b)

Fig. 7.1 A figure. (a) Graphical display, without annotation, a legend or title; (b) with badly placed labels that are too small and make it unclear what is being annotated; (c) the same figure is much clearer with annotation, a legend and a title.

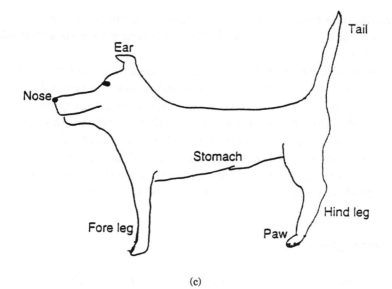

(c)

Fig. 1.1 A dog. The essential external anatomy of a dog is shown; the stomach is an internal organ.

Fig. 7.1 *(Continued)*

Figures can be:

* hand-drawn
* photographs
* images that have been scanned into a computer
* images created in a computer program
* printouts of visual data stored in a computer file

Figures drawn by hand

You might want to make line drawings, for example, of a piece of apparatus or an experimental subject. Unless you are an extremely good artist, it is best to keep these figures as simple as possible. If you need to trace a line drawing, try to avoid tracing paper. You will get a much neater effect using normal A4 paper with a light box—a transparent drawing surface, lit from below, which shines light through the image giving an outline that can be traced easily.

If you are drawing by hand, especially if you have to make a large number of figures, go to your nearest art shop and invest in a set of draftsman's pens. These give clear, even, dark lines that are ideal for reproducing with a photocopier. Do not use a pencil or a pen that is likely to smudge. Unless you have a particular need for colour use black ink, as other colours do not photocopy so well. If you make mistakes, use white correcting fluid to get rid of them—build up several thin layers to cover the mistake, if you use too much at once you will create unsightly blobs on your figures.

Photographs

Photographs should be clear and in focus, and only include information that is of use to your reader. If, for example, you are photographing a piece of equipment, do so against a neutral background and do not have unnecessary details intruding into your figure. If you find extraneous data or detail in your photographs, get rid of it; similarly, if you have a small image surrounded by a sea of background, get rid of the background, it does not tell the reader anything. If you have to edit by hand use either a scalpel and ruler, or a guillotine to crop your photographs. Do not use scissors as it is very hard to get a neat finish with them. If your photograph does not clearly illustrate what you want it to, consider using it alongside a line drawing to highlight the important details.

Images which have been scanned into a computer

Most images (including photographs of raw data, line drawings, and computer traces) can be scanned into computer files. Once they are in such a file you can use drawing or graphics programs to arrange the images so they show the important information clearly. In some cases, such as with high resolution photographs, it may be better to use the original image as scanning can cause a loss of quality and resolution.

Figures created in a drawing or graphics program

There are many drawing and graphics programs and most are fairly easy to use, at least to produce simple images (see *Chapter 12: Resources*). It is worth spending a little time playing with these programs and learning how they work, they will allow you to produce clear neat images that can easily be edited and rearranged if necessary.

Printouts of visual data stored in a computer file

Many types of data are gathered on computer: collections of readings (such as measurements of optical density or circular dichroism), traces and scans (for example, from FACs analyses and seismographs), and images (such as those produced by confocal microscropy). These data can be printed out as visual images for inclusion in your thesis or dissertation.

Manipulating Images

Manipulation of images is easy, particularly with the aid of a computer program, and is often used to enhance images—they should not be used to alter data. Care should be taken with manipulating images as there is a danger of giving a false impression of the experimentation, which could be construed as fraud.

Annotating Figures

Annotation simply means adding explanatory labels, such as arrows, numbers or letters, onto the figure to indicate special features that you refer to in the figure legend or text. Most figures need to be annotated. When adding annotations make sure they are large enough to be easily seen, but not so large they overwhelm the image. Letters and numbers in 10 or 12 point type are usually all right, providing they are in a clear font. Do not italicise the letters or put them in bold, unless that is the correct convention; for example, gene symbols are italicised. You can use arabic

(1, 2, 3, 4), or roman (I, II, III, IV) numerals, and either capital (A, B, C), or lower case (a, b, c) letters. Lower case letters usually look neater. Keep your annotations clear and simple and place labels as close as possible to the objects they refer to. Do not overdo your annotations or you may clutter the figure.

It is best not to write annotations by hand as it is very unlikely that you will achieve a neat result. Use printed or typed labels, Letraset, or a similar type of transfer. If you are using labels, stick down their edges neatly so they will not cast a shadow when the figure is photocopied. Either glue them or use matt sticky tape that will not reflect the light back into the photocopier and spoil your figure. Keep more than one copy of each figure in case you make mistakes. If you are working with a light background then use black symbols. If you are working on a dark background then use white symbols. Keep the type and size of font consistent, and if using Letraset make sure you have enough for all your figures; you may not be able to buy the same type again if you run out.

If your data are in a drawing or graphics file on your computer, then use the program to annotate the figure. Choose a font that is easy to read and use it consistently. Data generated directly from computers can sometimes be annotated before printing, but this is not always possible. It may have to be scanned into a drawing program for annotation, or annotated by hand after it has been printed.

Annotating a large number of items in one figure

If you have to label a large number of items in one figure it is usually clearer to the reader if you give a name to an item rather than using a letter key that needs to be explained in the legend. If necessary your labels can be vertical, rather than horizontal. Generally speaking, keys are a nuisance for the reader, but if you do not have room to write the name, then choose your letter key sensibly. For example, if you are electrophoresing RNA samples from different tissues, labelling them B, H, L, for brain, heart, leg, will make it easier for the reader to follow what is going on in the diagram than if you label them A, B, C. When annotating figures with letters, make sure that all the letters align if they need to and are all on the same horizontal line and the same vertical line as appropriate (see Fig. 7.2).

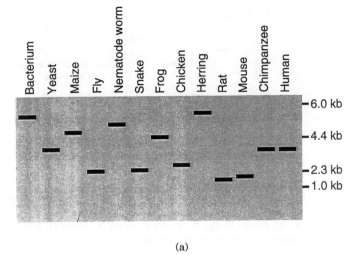

(a)

Fig. 3.2 Autoradiograph of DNA samples hybridised with *Ctbp* cDNA. DNA samples shown above were digested with *Eco*R I, Southern blotted and hybridised with the *Ctbp* cDNA, pNK1.

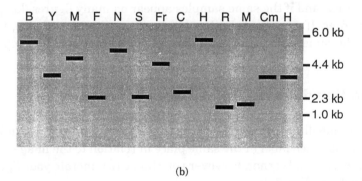

(b)

Fig. 3.2 Autoradiograph of DNA samples hybridised with *Ctbp* cDNA. DNA samples shown above were digested with *Eco*R I, Southern blotted and hybridised with the *Ctbp* cDNA, pNK1. (B, bacterium; Y, yeast; M, maize; F, fly; N, nematode worm; S, snake; Fr, frog; C, chicken; H, herring; R, rat; M, mouse; Cm, chimpanzee; H, human.)

Fig. 7.2 A figure containing many samples. (a) The samples are labelled with vertical annotation, and a scale is included; (b) the samples are labelled with a key that clearly indicates which is which; (c) the samples are labelled with a key that is non-obvious, the letters of the key do not align on the figure and there is no scale.

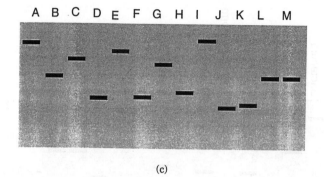

(c)

Fig. 3.2   Autoradiograph of DNA samples hybridised with *Ctbp* cDNA. DNA samples shown above were digested with *Eco*R I, Southern blotted and hybridised with the *Ctbp* cDNA, pNK1. (A, bacterium; B, yeast; C, maize; D, fly; E, nematode worm; F, snake; G, frog; H, chicken; I, herring; J, rat; K, mouse; L, chimpanzee; M, human.)

Fig. 7.2   (*Continued*)

If a number of related samples appear on one figure, put them in a logical order, and if the same samples appear on many figures, keep them in this order. If you show similar data in more than one figure, try and keep to the same scale within each figure, and the same size for each figure.

Annotating different figures within one figure

If you are labelling different portions of the same figure, A, B, and C, for example, put each label in the same place in relation to the image, such as in the top left hand corner. However you choose to annotate your figures be consistent (see Fig. 7.3).

Size and Scale

Tell the reader as much as is necessary about the size of the objects in your image. Either label each object with its size, or provide a marker or scale against which each object can be compared. In Fig. 7.2(c) it is impossible to tell the size of the DNA fragments in the diagram as no scale has been provided.

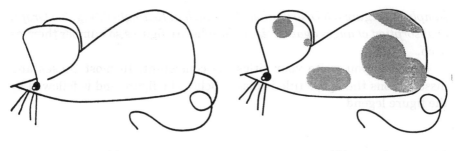

Fig. 4.2 Different types of mouse (a) a white mouse; (b) a light grey spotty mouse; (c) a dark grey spotty mouse; (d) a black mouse with white ears.

Fig. 7.3 A figure made up of different components. Each part of the figure is labelled, (a), (b), (c), (d), and the labels are in a clearly readable font and are positioned in the same place in each section.

Figure Titles

The figure title provides a heading for your figure, which will be listed in your Table of Figures at the front of your manuscript (see *Chapter 9: Deciding on a Title, and Planning and Writing the Other Bits*). Like all other headings it should be concise and descriptive; give the figure number and just one short sentence telling the reader what is there. The figure number can be written either *'Figure'* or *'Fig'*. Put the title in bold if you wish to. If you have a diagram you could write *'Figure 1.1 Diagram of ...'*, if it is a photograph you could put *'Fig. 5.3 Photograph of ...'*. Most people prefer to simply state the subject matter, *'Figure 7.6 Mass spectrogram of sample A', 'Figure 8.3 Distribution of landfish in Kellington', 'Fig. 2.5 Graph of absorption changes against time'.* Graphs are figures so number them as such.

However you write your titles, be consistent. In most theses and dissertations the figure title is placed below the figure and is followed by the figure legend.

Figure Legends

Figure legends give the reader a detailed description of what is in the figure. They explain the contents of the figure and where any data came from. Use the figure legend to describe interesting features that you have annotated. If you have had to label multiple samples with a key, then use the figure legend to explain what the key represents. Legends also allow you to cite references that are relevant to the figure. There are no rules about the size of figure legends, but very long legends are inconvenient to read. Remember to spell-check figure legends, and make sure they actually describe what is in the figure.

Do not reiterate the contents of a figure legend in your text, the two should be telling the reader different things: the figure legend is a detailed description of the figure, the text discusses the important highlights.

Black and White or Colour?

Generally there are no rules regarding the use of colour in theses and dissertations. Some of us do not use it because we do not have access to a colour printer. If you do have access to one use the colours cautiously. Colours can make a figure clearer, but if you go over-the-top it will look more like a psychedelic album cover than a scientific figure and will be difficult to follow.

Reproducing Figures

You will probably have to make copies of your figures. Figures can be photocopied, or if they are in a computer file they can be printed either onto normal A4 paper or onto photographic paper depending on the quality of the reproduction required. If possible figures should be reproduced after they have been annotated as you do not want to have to annotate the same figure a number of times.

Graphs

A graph is a type of figure that provides an easily understood summary of numerical data. Graphs get the message across more immediately than if the reader has to scan through a table of numbers, and they are extremely useful for showing the distribution of data, highlighting trends, and summarising relationships.

There are many easy to use computer programs that draw beautiful graphs from tables of data. It is worth spending time learning how to use these programs if you have access to one. Refer to the on-line help or manual for further information about producing different types of graph (see *Chapter 12: Resources*).

Here are guidelines for creating by hand the simplest types of scatter graphs, line graphs, histograms, bar charts and pie charts. If you are unsure which type to use for your analysis, consult your tutor or supervisor. Many readers of this guide will be producing considerably more complex types of graphs, such as those with non-Cartesian co-ordinates; your supervisor is your best guide for help in creating these graphs, but the basic guidelines are the same.

Some Basic Guidelines

Creating graphs

When planning a graph, think about what the final figure will look like— the simplest graphs are usually the most effective.

If you are not working in a graphics program, use squared paper and a sharp pencil for drawing your graph. You can go over pencil with ink later so it photocopies well. Draw all straight lines with a ruler. Clearly label what values you are plotting and the units you are using. Choose a scale that allows you to fit the graph onto one side of a sheet of A4 paper.

The horizontal axis is known as the x axis and the vertical axis is known as the y axis. Usually, values below the x axis and to the left of the y axis are negative. Generally, values that are chosen in advance in an experiment, such as time intervals, are plotted on the horizontal, x axis; these are known as independent variables or control variables. Variables that are measured at these points, such as optical density, are plotted on the vertical, y axis; these are known as the dependent variables or response variables.

The x and y axes do not need to cross over at zero, providing you clearly label the value of each axis where they intersect. For example, if the data you are plotting range from values of 1,550 to 1,650, there is no point starting this axis from zero as all of your data will be pushed into one small corner of the graph; in this case you could put a value of 1,500 or 1,550 at your origin (see Fig. 7.4).

Points plotted on a scatter graph or line graph are usually represented by a small cross centred over the data point or a dot surrounded by a small circle, making it clear where the dot is.

For some graphs it is appropriate to include error bars with your data points. Ask your supervisor for guidance if you are not sure whether or not to include them. If you do include error bars make it clear in the figure legend what they represent.

Annotating Graphs

Just like any other figure, graphs need to be annotated so the reader can interpret them, but try to keep annotations to a minimum, otherwise graphs tend to look very cluttered. You should:

Label what is represented by each axis

Use simple labels of just a few words maximum for the label, such as 'Time' or 'Weight', and use standard abbreviations. Do not use capital letters, italics or bold lettering for your labels as they tend to look crude or fussy.

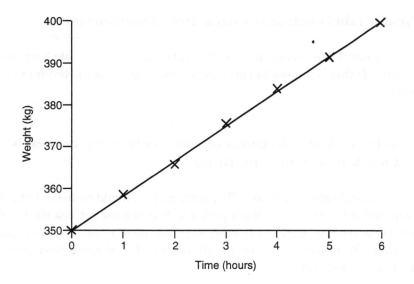

(a)

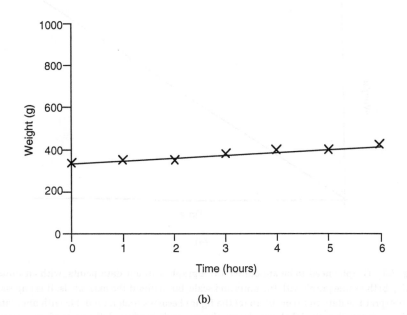

(b)

Fig. 7.4 Choose the appropriate scale for presenting your data. The scale in (a) shows the data well; that used in (b) is not a sensible choice.

Clearly label each axis with a unit of measurement

Use standard abbreviations and SI units ('m', 'kg', 'cd', etc.) wherever possible. If this is impossible provide an explanation in the legend to your graph.

Clearly mark the divisions of your scale along each axis, and mark in sensible multiples

Provide a scale along each axis. The scale marks should lie to the left of the *y* axis and below the *x* axis. Work with a scale that does not look cluttered— you do not need to mark every unit, often it is neater and as clear to mark the units in multiples of 5 or 10. We have included some good and bad examples in Fig. 7.5.

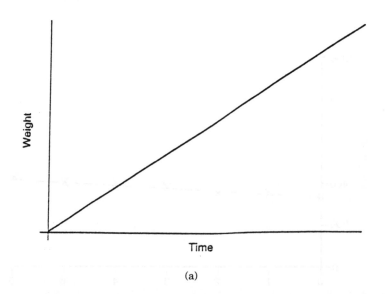

(a)

Fig. 7.5 Graphs need to be annotated. (a) A graph, without data points, without units or scale; (b) the same graph with the units and scale, but without the axis labels. It is impossible to interpret the data in (a) or (b). In (c) the figure becomes understandable with annotation: each axis is labelled, the labels are close to the axis and are centred, the units of measurement are given and the scale is shown clearly.

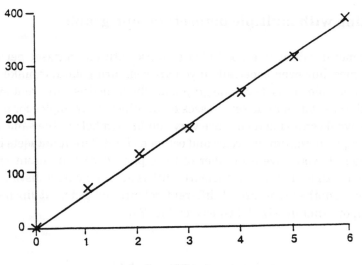

(b)

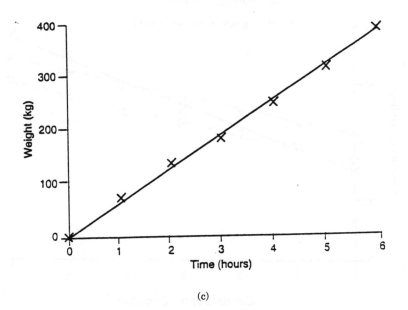

(c)

Fig. 7.5   (*Continued*)

Working with multiple datasets on one graph

As a general rule it is a good idea to work with one dataset per graph. Sometimes, however, especially if you are comparing data, it makes sense to plot multiple datasets on one graph, in which case use different symbols for each set—dots for one set, crosses for another set, triangles for another, etc. To avoid confusion do not use very similar symbols in the same graph, for example, using small circles and octagons for different datasets is not a good idea. If you have a number of lines plotted with different symbols make sure that they are sufficiently different to be easily distinguished from one another (the use of different colours for different lines can be helpful, or different sized dashes (see Fig. 7.6)).

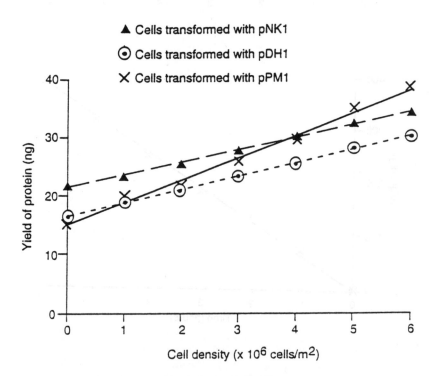

Fig. 7.6   Plotting multiple datasets on one graph. Plot each dataset with a different symbol and if possible each line in a different format (use colour if this helps clarify your graph). Label each line, but if this would make the graph messy, give a key as to what the symbols represent, as above, or explain them in the figure legend.

Sometimes it is better to produce more than one graph, rather than superimpose many datasets. If you have room on one page, you can keep your x-axis constant, but you can lay however many y-axes you need against it, clearly separated, but in one column. Beware of making such a complicated graph that it is undecipherable.

Working with multiple graphs

If you are showing similar data on more than one graph, try and keep to the same scale so the reader can easily compare them.

Figure Title and Legend

As with all other figures, graphs should be numbered (Figure 3.8 ..., Fig. 3.8 ...) and titles should be concise and simply state the contents of the graph. An explanatory legend may be useful, for example, for telling your reader where the original numbers are that were used to make the graph, perhaps in a table or in an appendix.

Scatter Graphs

Scatter graphs present the relationship between two different types of information plotted on a horizontal, x, and vertical, y, axis. You simply plot the point at which the values meet, to get an idea of the overall distribution of your data. Figure 7.7 shows a scatter graph.

Line Graphs

Line graphs present two different types of information plotted on a horizontal, x, and vertical, y, axis, where there is a sufficient relationship that you can draw a straight or curved line to join the points together.

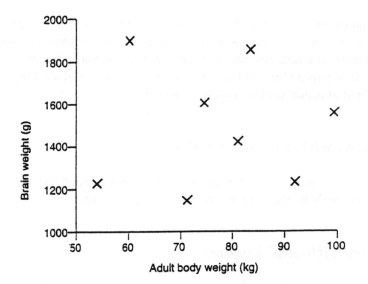

(a)

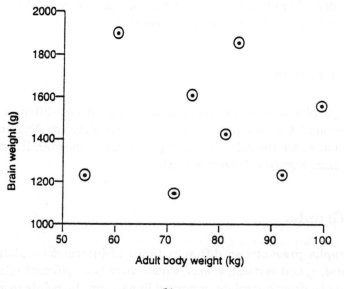

(b)

Fig. 7.7   A scatter graph. (a) Data points marked with crosses or alternatively (b) with encircled dots.

Drawing straight lines

The points on a straight line graph rarely fit the line exactly, so you need to draw the best fit, the best straight line, from what you have plotted. Different departments have different biases about how this should be done, so ask your supervisor for advice.

Drawing curved lines

If your points follow a curved line try to make your curve as smooth as possible. Make a photocopy of your plotted points and try drawing your curve on the copies, as many times as you like until you are satisfied with the result.

Interpolations and extrapolations

Using the line on a line graph to find out values that you have not measured, but which lie between the plotted points on the graph is called interpolation. Making estimates of values that lie outside the plotted points on a line graph is called extrapolation. If you are estimating values by drawing an extrapolated line on your graph then make this clear, for example, if you have been using a solid line for the data you have, use a dotted line for your extrapolation, and tell the reader what you are doing in the figure legend (see Fig. 7.8).

Histograms

These are useful for continuous data and percentages. When drawing histograms clearly state what you are plotting and the units and scale, and make the columns of equal width—this looks better, and for certain calculations you may wish to take into account the area covered by each column. In a histogram the columns touch, to indicate the data are continuous (see Fig. 7.9).

Do not spend ages creating amazingly intricate 'three-dimensional' graphs if you have simple data to present. The more complex a graph the more difficult it is for the reader to interpret.

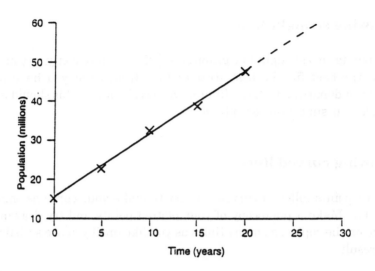

Fig. 7.8   Indicate where you are extrapolating. The dashed line is an extrapolated line.

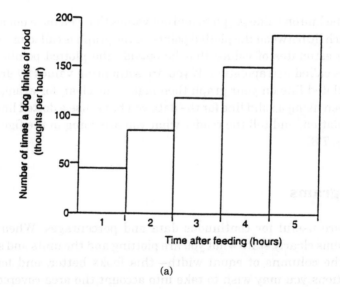

(a)

Fig. 7.9   A histogram. (a) The data are presented as a simple histogram; (b) the same data, but this time with a '3-D' effect, which is confusing because readers may think another value is being measured in the third dimension.

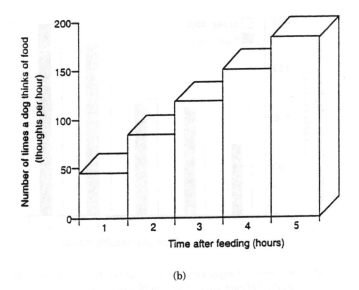

(b)

Fig. 7.9 (*Continued*)

Bar Charts

These can be useful for displaying values gathered for two or more data sets (and therefore the columns between each value on the *x* axis do not touch). Different columns represent different data sets and so you need to provide a key for the reader, and clearly state what you are plotting (see Fig. 7.10).

Pie Charts

Pie charts are a simple way of showing proportions and percentages. Again they need to be fully annotated so it is clear what each 'slice' of the pie represents and what proportion or percentage of the whole it represents (see Fig. 7.11). Sometimes people like to shade the individual sections of a pie differently, but this can be messy if it is divided into many parts.

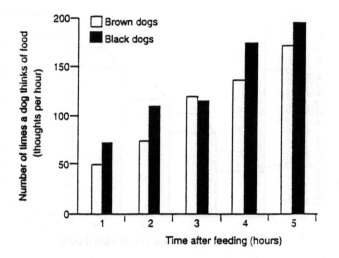

Fig. 7.10   A bar chart. The label explaining the entries ('black dogs, brown dogs') should be easy to read and the shading of each column should photocopy well.

Fig. 4.1   A dog's dinner. The different components of a dog's diet.

Fig. 7.11   A pie chart. This is useful for showing percentages. Shade the 'slices' if it clarifies what you are trying to show.

Tables

Tables allow you to present a large amount of data that is relevant to more than one sample. Unlike graphs, in a table you show the exact values of your data. You need to make your tables as easy to read as possible. Decide what you want the table to say, and then decide on the best way to say it.

Some Basic Guidelines

Creating tables

For simple tables with few entries, it will probably be easiest to use your tab key to create the columns. If you are unsure about the use of tab keys see *Chapter 12: Resources*. The tab command will let you align your text to the left, right or centre. Normally tabs align to the left, for example:

mouse	cheese	bacterium	milk
cat	mouse	cheese	bacterium
dog	cat	mouse	cheese
cow	dog	cat	mouse

or,

1998	716	298	1678
14	9	19	59
25	10	1960	1
304	1109	1	782

This may be useful for certain tables, for example, where the entries are words. However, if you are using your table to present numbers, it is best to align to the right, as this makes it easier to compare and process the numbers:

1998	716	298	1678
14	9	19	59
25	10	1960	1
304	1109	1	782

You can also align to the centre, but generally this is of little use for scientific theses or dissertations:

1998	716	298	1678
14	9	19	59
25	10	1960	1
304	1109	1	782

If you have numbers with decimal points, the point should be positioned in the same place in each column, as in the example below. If you have values of less than 1, always put a 0 in front of the decimal point so that it is your figures read '0.9', '0.043' etc., not '.9', '.043'.

1998.01	71.60	2.98	16.78
1.42	9.00	19.00	0.59
0.25	0.10	19.60	0.01
30.45	110.96	1.67	78.29

For more complex tables it is easier to use the 'Tables' command in your word processing file, or to use a spreadsheet program, which will allow you to set up and edit tables quickly and easily (see *Chapter 12: Resources*).

If you have a zero reading enter it as 0. Different disciplines have different conventions for what to write if you have no reading—most people either leave a blank space, or put in a code such as 'ND', which should be explained in the legend ('ND not determined'). Do not use a dash '-' as this may be confused with a negative sign '-'. Check what your supervisor prefers.

Table layout

Think about the order of columns and rows; place any data that can be compared in adjacent positions. If you have related numbers, put these in columns so that people can easily scan up and down, rather than from side to side. Arrange columns and rows of related data together in your table, in a logical order.

Enter your information in an appropriate form using correct scientific notation. Clearly state what units, if any, you are using and use SI units and standard abbreviations. Add in error statistics for numerical values if necessary.

Annotating Tables

Tables need annotating, but keep the annotation as simple as possible so that your tables do not look cluttered. Each column and row should be labelled explaining its contents and giving units of any values you are entering. If you have long headings, use two or three lines to fit them into the space above the column, or beside the row. It is not necessary to draw lines between columns and rows, doing so often makes the table look fussy. There is no reason why every column should be as wide as every other column, or every row should be as deep as every other row. Space rows and columns according to the amount of text they contain.

Name	Colour	Age (years)	Ear length (cm)	Number of dog biscuits eaten
Bonzo	black/brown	2.6	10	45
Rover	brown	5.7	15	78
Felix	white	3.0	6	2
Henry	brindle	8.4	12	83
Toby	black/white	15.1	20	169
Groovy	orange	10.9	17	50
Tim	black	15.2	14	39

Numbers and Titles for Tables

Each table should have a number and a title which clearly and concisely indicate the contents. Keep it as simple as possible. It looks neater if you centre the title and avoid using capital letters or italics. The table title is listed in the List of Tables at the front of your manuscript (see *Chapter 9: Deciding on a Title, and Planning and Writing the Other Bits*).

Table 3.4 Food stimulus response in dogs

Food	Tail wag (per minute)	Bark (per minute)	Jump (per minute)
Dog biscuits	120	83	12
Dog food	102	52	8
Doors	68	34	7
Footwear	116	72	9
Grass	1	1	0
Jumpers	9	5	2
Sausages	150	49	11

Footnotes for Tables

If you need to add footnotes to your table it is best not to use numbers as symbols because these may be confused with your data. Use asterisks, stars and so on, $^{*\,\dagger\,\S\,\P}$. It is best to put them in superscript so they stand out.

Inserting Figures and Tables into a Thesis or Dissertation

Refer to the figure or table in the text; do not just stick it in and hope the reader realises what it is doing there. Check you have referred to each figure in the text, and that there is a figure present for each one that is referred to.

Figures and tables are either printed with the text, or stuck into the manuscript after it has been printed

Figures can either be inserted into an appropriate sized space in your text or be printed onto a separate sheet of A4 paper which is then added to the text. If possible put your figures and tables in 'upright' so the reader will not have to turn your thesis or dissertation on its side to read them. If this is not possible put them in so the bottom of the figure is always at the right hand side of the text. If you try to put the bottom of the figure on the left hand side, the figure legend will probably be lost in the binding.

See Fig. 7.12. All pages containing figures should be numbered.

Importing figures, graphs and tables into word processing files

Most word processing programs allow you to import figures and tables from other types of programs, such as graphics programs or spreadsheets. This is an easy way of incorporating figures into your text. Look at your program manual and the help option for further instructions. Check carefully, on a printout, that no glitches occurred when the figure was imported.

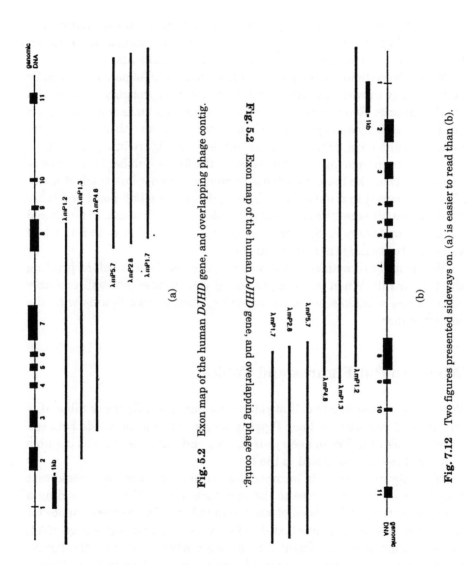

Fig. 5.2   Exon map of the human *DJHD* gene, and overlapping phage contig.

(a)

Fig. 5.2   Exon map of the human *DJHD* gene, and overlapping phage contig.

(b)

Fig. 7.12   Two figures presented sideways on. (a) is easier to read than (b).

Adding figures and tables to the text after it has been printed

First of all you need to ensure you have enough copies of each figure. For example, if you have to submit three copies of your thesis or dissertation make at least five copies of each figure in case of accidents.

With a large figure it is a good idea to print it onto A4 paper so it can be bound neatly in with the rest of your text. You will probably need to trim smaller figures. Remember to use either a modelling knife and a ruler, or a guillotine (which your department should have). You can either place several images on one page, or incorporate a single small figure into your text.

As you are preparing your figures write the figure number (as it appears in your text) on the back of each one so that you do not lose track of them. Use a soft pencil to number your figures, if you use a ball-point or felt-tip pen it can easily show through. Put the copies of your figures into numbered envelopes or folders so they are clearly organised and easy to find when you come to sticking them into your text.

We strongly recommend that you prepare the figures before printing the text in full. When you are preparing your figures you will probably come across mistakes in your numbering, such as two figures with the same number.

Where to Put Figures and Tables

When you are reading text that refers to a particular figure or table, it is irritating if you have to keep flipping backwards and forwards between pages to look at it. Try to keep your figures and tables as close as possible to where they are discussed in the text.

If the figure or table is incorporated within your word processed text file, it is easy to put it close to the appropriate text. If you are adding a figure after the text has been printed, remember to leave enough space for it in the final layout of your text. If a figure takes up an entire page of A4, put in a page break and print a blank page where you want the figure; insert the figure in place of the blank page. This way you will not disrupt your page numbering.

Including Figures and Tables on Computer Disc, Video Tape, CD, and Audio Tape

Sometimes data need to be presented that cannot be shown on paper, for example, data on video tapes, or very large bodies of data that need to be submitted on a floppy disk or CD-ROM. If you have to include a small flat object such as a CD or computer disc, you can either make—or if you are having the thesis bound ask the binder to make—a pocket at the back of the thesis. If you have to submit a bulkier object such as a video, ask your department or university if there are any rules concerning its submission. Make sure that any objects you submit as part of your thesis are clearly labelled. The labelling must include your name, your department, the date and which degree you are doing.

Common Mistakes

- Not referring to a figure in the text
- Referring to data in the text but not illustrating it with a figure
- Spelling mistakes in the figure legends
- Not adding scales or scale bars to figures
- Using cluttered graphs with too much data
- Using cluttered tables with too much data

Key Points

- Decide which figures and tables you need to include when you write your thesis plan
- Prepare a draft of each figure and table
- Annotate figures and write a short legend
- Label each axis of a graph and add units and scale
- Label the back of each figure before inserting it into the final copy of your text; then check the numbers are correct for the figure legends

Chapter 8

DECIDING ON A TITLE, AND PLANNING AND WRITING THE OTHER BITS

Fig. 8.1 Choosing a good title can be difficult.

The Title of Your Thesis or Dissertation

A short self-explanatory title is best. If it is too short the reader will not be able to tell what the thesis is about. If the title is too long the reader will have lost interest by the time they finish reading it. We will look at three possible titles for a thesis that addresses the question *'What are the courtship and reproduction rituals among landfish along the M4 motorway?'*

Landfish

Whilst this is a very compelling title, it is more suitable for a novel than a thesis. It does not make clear what specific question the research project was asking, and implies that the thesis covers the complete field of landfish biology. This title is unhelpful to potential readers and in an oral examination the writer could justifiably be asked any question about landfish, rather than the one aspect of their behaviour that has been studied.

An Ethological Analysis of Sexual Behaviour—Courtship, Mating, and Reproductive Rituals and Rites—among Male and Female Landfish (*Gallus fritos*) on the Embankments and Adjacent Regions of the M4 Motorway between Junctions 5 and 25, Encompassing and Including Areas Between Wokingham and Chipping Sodbury

This is far too long and detailed. It is repetitive, unwieldy, and difficult to read. Cultivate a concise yet descriptive, efficient and accurate style in all your writing.

Courtship and Reproduction Rituals among Landfish Along the M4 Motorway

This is just right—the title lets the prospective reader know what the thesis is about without going into redundant detail.

It can be difficult to come up with a good title. If you are lacking inspiration then try re-writing your aims. For example, if your research addressed the question *'How is the cognitive development of 3 to 6 month old babies affected by external stimuli?'*, your title could be *'The effect of external stimuli on the cognitive development of babies aged between 3 and 6 months'*. If you need more examples look at good recent theses in your field. Do not worry too much about sounding dry or stuffy as long as the title does its job. It is best to avoid redundant phrases such as *'A study of …'*, or *'An investigation of …'* People should have realised it is a study or investigation of something by the fact it is being presented as a dissertation or thesis.

The Other Bits

If you look at a recent thesis or dissertation in your field you will see that it does not simply start with page one of the Introduction. There will probably be an opening section made up of at least a Title Page and Table of Contents, which are there to help the reader navigate around the thesis or dissertation. You will need to include all or some of the following sections, and although there are usually no rules about the order in which they are presented, the layout below is often used:

Title page
Abstract
Dedication
Acknowledgements
Table of Contents
Table or List of Figures
Table or List of Tables
Table or List of Appendices
List of Abbreviations
Main text of the thesis including References
Glossary
Appendices
Published Papers

You might use one or two, or all, of these sections in your thesis—it depends on the conventions of your field and the nature of your thesis or dissertation; we recommend you see a good recent thesis in your department for guidance.

Many people create a separate computer file for the sections at the front of the thesis, which precede the main text. If you do so you can edit it without altering the page numbers in the main text. It is a good idea to distinguish this section by numbering the pages with roman (I, II, III, ... , etc.) rather than the arabic numerals (1, 2, 3, ... , etc.). Your word processor program will allow you to insert page numbers in roman numerals.

Title Page

Each institution will have specific requirements for what should go onto the title page of a thesis or dissertation. Check what these are in your department. Generally the title page requires your title, your name, your affiliation (department and institution), and a statement indicating for which degree you are entered, for example:

> Submitted in partial fulfilment of the requirements for the degree of Bachelor of Science of the University of West Cheam.

Some title pages will also include the month and year in which the thesis is submitted. Figures 8.2 and 8.3 show typical title pages for a BSc thesis and a PhD thesis that are being submitted to the University of West Cheam.

Abstract

In many ways your Abstract (or Summary) does the same two jobs as the trailer for a film or the blurb on the back of a novel—it says what your thesis is about in language that general readers can understand and it highlights the juicy bits that will make them want to read it. This is probably the first time you have had to summarise your entire project, but do not panic—you know your research better than anyone.

The length of your abstract

Although there are usually no rules as to the length of your Abstract, it is a good idea to keep to just one side of paper. The Abstract is written for two audiences, the examiners and people researching in your area who may be interested in your results; they both need a concise and easily read *précis* of your work. By confining yourself to one side of paper you will necessarily keep to the main points, and satisfy both audiences.

Planning your abstract

Although your Abstract is only one side long, plan carefully before writing it. Note the key points you wish to include. Start broadly by introducing your field of study, narrow down to describe your experiments, then assess

Mapping Genes Associated with Neurodegenerative Diseases on Landfish Chromosome 4

Adrian Thaws

Submitted in partial fulfilment of the requirements for the degree of Bachelor of Science of the University of West Cheam June 1999

Department of Molecular Neurogenetics, University of West Cheam , Tennis Pavilion Buildings, Nonsuch Park, Surrey

Fig. 8.2   Title page from a BSc thesis that is being submitted to the University of West Cheam.

The application of mathematical models to the problem of moving parts

Stephen Tyler MA

**A thesis submitted for the degree
of Doctor of Philosophy,
University of West Cheam**

**Department of Mechanical Engineering,
University of West Cheam,
Salisbury Avenue, Cheam**

Fig. 8.3   Title page from a PhD thesis that is being submitted to the University of West Cheam.

the importance of your experiments and their wider significance. You could approach the task by taking notes in the following order:

- Field of study: Introduce your field then note your specific area of interest.
- Aim: What have you been trying to do or show and why?
- Experimental system: How did you address your aim?
- Results: What have you achieved, did everything go as planned?
- Discussion: What importance does your work have in relation to the rest of your field of study?

Keep your notes simple and only pick out the key details.

Writing your abstract

Your Abstract should allow another scientist in your discipline who is not a specialist in your topic to get an idea of what you have done, so try to write in accessible language and avoid specialist terms as far as possible. The reader should be able to tell immediately from your Abstract if your thesis is of interest to them. Avoid using headings and sub-headings as they will clutter such a short piece of writing rather than make things clearer.

The contents of your Abstract should mirror those of the main text. As a very rough guide: your introductory paragraphs should be 20–30% of the whole; your methodology paragraphs should be 10–20%; your results paragraphs should be 35–45% and your discussion paragraphs should be 20–25%. Do not include figures or tables in your Abstract.

Dedication

Most people do not bother with a dedication as well as an Acknowledgements section, but you might just wish to put in a short one sentence dedication to someone or something special, such as your parents, partner, or dog.

Acknowledgements

This section is the most informal part of your thesis. It allows you to thank everyone who has helped you during your research project: your lab mates and supervisor, parents or partner, friends who cooked you meals and bought you drinks. You may also wish to include people in your local engineering workshop, administrators or other people if they have been helpful. It is a good idea to be generous with your acknowledgements: this is the only bit of your manuscript that everyone is going to read and people remember whether or not they were mentioned.

Table of Contents

This is exactly what it says it is. We have a Table of Contents at the beginning of this book—a list giving the page number of each chapter, heading and sub-heading. You can create a Table of Contents in two ways either by automating the process using an 'outliner' on your word processor which tags each sub-heading and automatically creates a Table of Contents (see *Chapter 12: Resources*), or you can create the Table of Contents yourself. The easiest way to do this is to (1) create a new file for the Table of Contents, (2) make a duplicate copy of your finished thesis, (3) go through this copy and paste headings from it, into the Table of Contents file, making sure you note the page number for each entry as you go along. Never never do this working from the original version of your thesis, in case of a word processing disaster.

Table or List of Figures

The Table of Figures helps the reader to find the figures they are looking for. Look through the final version of your thesis and list the figure number, the figure title, and the page it is on. Graphs and charts are figures so include them in this table.

Table or List of Tables

You should have a separate table listing your tables, by number, with titles, and giving the page they are on.

Table or List of Appendices

You may have appendices at the end of your manuscript that present useful information that is too bulky or intrusive to include in the text, but which examiners need to know. The Table of Appendices helps the reader to find this information easily.

List of Abbreviations

Provide the reader with a list of all abbreviations you have used including all non-standard abbreviations, so that there is absolutely no room for ambiguity. Many everyday abbreviations we use around the laboratory are non-standard and will not necessarily be understood by someone in another laboratory, let alone by an examiner from another university. Although not strictly necessary, it is a good idea to incorporate the standard abbreviations, such as chemical or physical symbols, into your List of Abbreviations. Keep a list of these abbreviations as you write them in your text, and put them in alphabetical order and with explanations in one list.

Glossary

A glossary is a type of mini-dictionary of specialist words. You could include one at the end of your thesis; for example, anyone writing a thesis about a particular aspect of medical research might want to include a glossary of specialist medical terms. If you are going to include a glossary add words to it as you go along. Then sort them into alphabetical order with explanations.

Appendices

Appendices are the sections at the back of your thesis where you can put large amounts of information that would detract from the flow of your main text, but which are still necessary to include. Put data and information that the examiner does not need in order to understand your argument, but does need in order to check your argument, into appendices. In an appendix you could present, for example, tables of data, the detailed results of database searches, calibration measurements, computer programs, user guides for software, or lists of methods and suppliers. You could also include published information, such as well-known proofs, that is necessary to support your work, to save the reader having to look it up in a library. Put different types of information into different appendices. Give each appendix a heading and number, and list them in a Table of Appendices at the front of your thesis or dissertation.

Computer Programs

If you are including a computer program in an appendix, print it on the same paper as the rest of your thesis. It is not necessary, particularly with long programs, to include non-essential commands. Print in multiple columns, use single line spacing and a small, but readable font, to save space. The program may be so long that you need to include it on floppy disk or CD-ROM, in which case label these with your name, address, department and course, and ensure they are safely submitted along with the text of your thesis or dissertation.

Published Papers

If you have been an author on published papers that are relevant to your project, it is a good idea to include these papers with your text as they will help to show the examiners what you have achieved. Copies of the papers can be bound into the back of your dissertation or thesis. You may also include papers that are 'In Press' (accepted for publication, but not yet published) or that are 'Submitted' (currently being reviewed for publication), providing that you clearly state 'In Press' or 'Submitted' on the title page of the paper.

Common Mistakes

- Inaccurate titles
  Make sure your title really does describe the contents of your thesis. We came across one, otherwise excellent, PhD thesis suffering the indignity of having had its title slashed through with red pen. The author had been made to change the title even though there were only a few corrections in the main body of the thesis.
- Inaccuracy in the Table of Contents—missing out sections of text or writing incorrect page numbers

Key Points

- Choose a title that is focused and best reflects your Results
- Check and fulfil any regulations concerning your Title Page
- The Abstract is a short summary of your thesis—keep it to one side of a sheet of paper
- Create a Table of Contents for the reader, and if appropriate include other Tables for Figures and Appendices, etc.
- Make sure you have a List of Abbreviations

Chapter 9

PROOFREADING, PRINTING, BINDING AND SUBMISSION

You have already put a lot of hard work into writing your dissertation or thesis. The final steps, printing, binding, and submission, are fairly straightforward but still require close attention to detail. It is silly to spoil an otherwise good text by allowing mistakes in spelling and layout to mess it up.

Know Your Submission Date

If you are an undergraduate or masters student you will almost certainly have a deadline for submission set by your department or university—know this date. It is surprising how often people do not know the timetable of their course. PhD students do not normally have a set deadline for submission: they are usually allowed up to seven years from the beginning of their course—check with your local administrator for further information.

Whether you set your own deadline or have one set for you, draw up a realistic timetable including enough time for printing, sticking in figures, proofreading, and binding. This process can take a couple of days, possibly longer for a PhD thesis.

Fig. 9.1 Know your submission date.

Know How Many Copies You Need

For most undergraduate theses or dissertations you will be required to submit only one copy, for PhD theses up to six copies may be needed. Know what you need so that you can prepare sufficient figures and text.

Advance Organisation

Most people who have written a thesis or dissertation have experienced the frenzied last minute rush for a deadline. Generations of supervisors have said to generations of students 'Leave yourself enough time at the end'. Generations of students have ignored or forgotten the advice and had to spend the last week or so before their submission date sleeping at their desk, if at all, as they work through the night to get their theses finished. People find themselves printing the final version of the dissertation or thesis at 4 am only to discover that the printer runs out of toner, the glue for sticking in the figures goes missing, computers and scanners mysteriously stop working, and the entire departmental stock of paper runs out because all the other students are printing their theses at the same time. To avoid this nightmare, prepare in advance.

Advance Organisation of Figures

It is amazing just how long it can take to add your figures to your text, and just how muddled the process can become when you are tired. To streamline the final production of your thesis prepare your figures in advance. Figures and tables can either be added to your text by hand, or imported into the text file on your word processing program and printed with the text. Make sure you have any figures you will be adding by hand prepared and ready for insertion (see *Chapter 7: Figures and Tables*).

Preparing to Print the Final Version

Once your text and figures are ready and your thesis or dissertation is laid out properly, allow yourself enough time for some final checks. Spell-check everything—including figure legends and titles, and the annotations to figures and tables. Spelling mistakes will lose you marks, and, if you are submitting a thesis as a publication (for example, a PhD thesis), you will have to go back and make whatever spelling corrections your examiners request, before you can be awarded your degree. Remember to spell-check the References, which is a painful and time-consuming task because the program will probably query most of the names. Any misspelt names or titles could be spotted by your examiners. Check for contractions like 'it's' and 'isn't' (try doing a 'find' command for apostrophes in your text) and commonly confused words such as 'proceed' and 'precede' (see *Appendix 1: Easily Confused Words*).

Before you finally print your text, make sure it is properly laid out by using the 'print preview' facility of your word processor. This lets you see what each page looks like on your computer screen before committing it to paper. Check that the layout of the thesis is consistent—that your titles are all in the same font, and similarly, of course, *your text*. Also check that all the heading, table, and figure numbers are correct and that all your titles have text underneath them. Something that many people miss before printing is having a title on the last line of a page and the text starting on the next page, which looks awful. Check that you have remembered to put in page numbers and left enough space for any figures you are going to put in by hand.

Once you have looked at your text on computer, print one copy of your thesis. Check that everything looks consistent and is well spaced, and that you do not have blank pages or other unexpected mistakes. Now thoroughly proofread your text.

Proofreading

Proofreading your text means checking it for mistakes, (from grammatical blunders to minor typographical errors), before submitting it. Do not rely on your spell-checker, it will only pick out combination of letters it does not recognise as words. It will not pick up grammatical mistakes, words that have been used incorrectly, or words that have been missed out. Make sure all your numbers are correct; it is very easy to put a decimal point in the wrong place. If you are using equations be absolutely sure they are right. At this stage do not make major changes to your text—stick with what you have written.

It is possible to proofread the text yourself but much better to have someone else proofread it. They will find mistakes that you have missed. Choose someone who has a good command of English. If you are getting a friend or relative to proofread your text you could use any form of notation you like for the corrections, providing you both understand it. Once your text has been proofread make your final corrections. Do not rush this stage just because you are almost finished.

Printing Your Dissertation or Thesis

Prepare adequate stocks of paper, photographic paper, toner and any other supplies you may need (including strong coffee), so that the final stages in the production of your thesis go as smoothly as possible. It is also worth finding out where alternative computers and printers are, just in case you need them.

Before you print, clear away any old drafts of your thesis that may confuse you and any other debris: crusty old coffee mugs, the remains of your last three meals—anything that might spoil your handiwork. You may be tired and accident-prone and it is very easy to knock a cup of day-old coffee over your final draft or drop your beautifully prepared figures into a bowl of cornflakes just as the last page of your text is crawling out of the printer.

Fig. 9.2   Prepare to print your dissertation or thesis.

Check, again, that you have a stack of clean paper in the printer and the toner is not run down. You can now give the print command to print the final version of your thesis. It is sensible to print one copy and do a last minute check for any mistakes before printing further copies.

You will almost certainly have to print more than one copy. Many computers will collate multiple copies for you. On other computers if you give the command for multiple copies you will get multiple copies of each page and will have to collate them yourself. Know how your print command works so you do not have to spend hours collating by hand. Whether or not you collate by hand, make sure that each copy of your thesis or dissertation is complete and in the right order. Students have been known to submit copies with two chapter ones and no chapter twos.

Adding Figures to the Completed Text

While your manuscript is printing relax for a few moments, then prepare to stick in any figures that need to be added. Do not use sticky tape—it looks awful—use mounting-spray or glue-stick, which give a more professional finish.

Organise all your figures in numbered envelopes (see *Chapter 7: Figures and Tables*). Clear some clean space and get out the glue. When the copies of your text have finished printing and are all in order, stick in the figures. Start by putting a copy of your figure 1 into each copy of your thesis or dissertation, then do the same with the copies of figure 2, figure 3, and so on. Check the right figure is in the right place: if you have numbered each figure on the back in pencil, this should not be too difficult.

Take care not to use too much glue or else your pages will stick together. If you are using a spray glue then turn the figures face down on a clean piece of scrap paper, spray the glue, and stick in the figure. Throw away the piece of scrap paper and use a fresh piece for the next figure, otherwise you are bound to get glue all over the wrong side of your beautifully prepared figures and waste all the time you spent earlier. Whatever you do, do not simply plonk your figures down on your work surface and start spraying them. You will mess up your figures and you will mess up your work surface.

Once you have stuck in all your figures make a final check. Look through your text making sure the page numbers, figures, and figure legends are all consistent. Place each copy of the thesis in a folder and, if necessary, go to bed.

Including Your Own Publications

Copies of any of your own papers related to your research project can be bound into the back of your thesis or dissertation. You can include papers that are 'In Press' (accepted for publication, but not yet published) or that are 'Submitted' (currently being reviewed for publication), providing that you clearly state 'In Press' or 'Submitted' on the title page of the paper.

Binding

You now have a very valuable stack of papers. These need to be bound together so that none are lost and the thesis or dissertation is easy to read and handle. How your text should be bound will depend on your department and at what level you are studying.

You must check with your department or university about rules covering binding. For an undergraduate dissertation, for example, you could buy a plastic spine that slides along the left hand side of your stack of papers to hold them in place. A sheet of transparent plastic, such as the type used for overhead projectors, at the front and back of the text will make it look neater and help protect it. Some departments have access to binding machinery in their administrative offices which produce either spiral bound or glued theses. If such a service is available to you, find out in advance who runs this service and make sure that they are ready for your thesis, especially if you are going to be in a rush to get it submitted. This type of binding is usually fairly speedy, but if an entire class is waiting, you do not want to be last in the queue.

For higher degrees you are usually required to submit a hardback professionally bound text. Some institutions will allow you to submit a text in a temporary binding, so that corrections can be done easily; however, the text will have to be properly bound once these corrections have been made. If you have to submit a professionally bound text find out the name of recommended local binders. Your university or department should be able to provide you with a list. Professional binding can be expensive, but often binders will reduce the charges if you are not in a hurry and can wait for them to fit in your job when they have little else to do. Telephone the binders to check what they charge and where they are.

For the higher degrees such as MScs and PhDs universities usually have strict rules about the cover and title page of the thesis: what colour it should be, the font and colouring of the letters and what it should say—for example name, degree title, date. Many experienced professional thesis binders will know the rules for their local university or college, but you should make sure you know the rules yourself. Ask your supervisor or departmental administrator.

Submitting Your Thesis

Submitting your thesis is the easy bit. All you have to do is give it to the relevant authority for examination. The one thing you do have to be absolutely sure about is who and where the relevant authority is. The Finance Department will probably find your text very useful as a doorstop, but they are not authorised to award you with any academic qualification. Find out from your supervisor or administrator where you need to go and what their office hours are so that as soon as your thesis is bound you can hand it in.

What Happens Next

What happens next depends on the degree for which you are studying. For undergraduates and some master's courses, your dissertation or thesis grade might simply be part of your final mark. For doctoral students you may wait several weeks for an oral exam, a viva, in which your thesis is scrutinised by two experts (see *Chapter 10: And You Thought it Was All Over*).

Common Mistakes

- Not allowing enough time for the final stages of preparing a thesis or dissertation
- Not thoroughly checking the final version for errors
  Several years ago, a friend, who is now a professor at a major Californian university, was running very close to the submission deadline for his BA

dissertation and it looked as if he would miss the deadline. He had arranged with the friendly college librarian for her to bind the thesis for him. By midday of the last possible day he could submit the work he was still printing it and sticking in his figures. By 4.30 pm he had finished. He collected all his papers together, ran from the computer room to the library, up the steps and into the arms of the waiting librarian. She grabbed all four copies of the thesis, shoved them through the binding machine and handed the bound copies to our flushed and nervous hero. By this time it was 4.45 pm—the deadline was 5 pm. He then ran from the library to the submissions office, arriving at 4.55 pm. He had five minutes to spare and nonchalantly handed the four copies of his text to the administrator, who opened the theses to find all four copies had been bound down the right hand side of the page.

Key Points

- Find out about submission rules and dates
- Do not underestimate the length of the printing and binding process
- Prepare any figures that need to be stuck in before you print the final version of your thesis or dissertation
- Spell-check, proofread, and check the layout of a printout of your thesis or dissertation before printing the final copies
- Carry out one final check of copy of your manuscript before it is bound

Chapter 10

AND YOU THOUGHT IT WAS ALL OVER

Oral Examinations

There are two reasons you might have to take an oral examination, commonly known as a viva (an abbreviation of the Latin *viva voce* meaning by, or with the living voice). At BSc and MSc levels, you might have to take a viva if your exam results are borderline, for example, between pass and fail, or first and second. Alternatively, if you are writing a dissertation or thesis, a viva may be part of the exam process which everyone has to go through—depending on the institution and course. In this chapter we are only considering those vivas which concern theses or dissertations and are a set part of a degree course.

As with any other exam, you are more likely to pass if you prepare properly for the viva and hone your technique. The examiners are not there to make you fail. They will probably try to make the viva as enjoyable as possible both for you and for themselves. Some people actually enjoy their vivas—for a PhD student who has poured their heart and soul into a project for three years or more, the viva is probably the last chance to really discuss their work with experts who have studied it in detail.

Your examiners will be trying to make sure that you have really thought about your project, both its data and implications, and that you have an understanding of your field appropriate to the level of your degree. Just as you are more likely to convince your examiners of your professionalism if your thesis is well planned and well presented, you are more likely to convince them of your worthiness at the viva if you plan for it and present yourself well.

What to Expect in the Viva

All vivas have a similar format, whichever degree they are for. Your university or department will arrange a venue and the time. You enter the room and are questioned about the contents of your thesis and background knowledge. For a short BSc thesis this might last 10 minutes; for a PhD thesis it could take several hours. The examiners will tell you when the viva is over. It is a gruelling experience for both you and your examiners.

For most vivas you will be asked questions covering your project, your general field of research, and the background to your project. Questions about your project will be both theoretical (why you took certain approaches, why you developed certain strategies, what the principles behind certain techniques are) and practical (exactly how you carried out your experiments, and the detailed interpretation of your data).

The level of knowledge that you are expected to have will depend on which degree you are taking. PhD students have had at least three years more study than undergraduates and thus will be expected to know their chosen field in greater depth.

Preparing for Your Viva

Know your thesis

A cliché that is often used to make new seminar speakers feel less nervous is that '*you know your own work better than anyone*'. Like most clichés, it is true. Read through your thesis carefully several days before your viva. As you are reading, think what questions the examiners might ask. Note these questions on a sheet of paper and make sure you can answer them correctly. This will give you some idea of how much you really know. If you find yourself guessing or floundering on any of the questions, revise this area of your knowledge. Think about why you took certain approaches, and whether you would do the same again in retrospect. You should also think about what future work might come from your thesis.

Your viva is an oral examination so it is a good idea to practice answering the questions out loud. Your flatmates will have got used to your strange behaviour by now, so they will not think it unusual to hear you talking to yourself. If you can, get a friend to run through the thesis and ask you questions. You can also ask your supervisor to arrange a 'mock' viva, so that you get some feel of what the real thing is like. Not all supervisors will agree as it can be an enormous amount of work for the 'mock' examiners, who simply may not have the spare time, but it is worth asking.

Discuss your project with your supervisor before the viva. Ask them for comments and criticisms and tell them about your ideas and your response to likely questions; your supervisor should be a helpful sounding board.

Spotting your mistakes

If you have found errors in your thesis or dissertation, either typographical or in the data, make a note of them and bring a list to the viva. The examiners will appreciate that you have thought about your thesis and are honestly trying to produce the best possible piece of work.

Know your field

Your examiners will not confine their questioning to your thesis or dissertation; they will also examine, to some extent, your knowledge of the field in which you work. You can never know exactly what the examiners will ask, but you can feel prepared if you have considered the following questions:

- What research and what ideas led to the creation of your field?
- What research and what ideas led to the creation of your project? What questions has your project answered?
- Have similar projects to yours been undertaken before; what have these told us about your area of research?
- What have similar projects told us about the techniques you used?
- What are the implications of your area of study for other fields, for example, how does biotechnology affect environmental sciences; are any ethical issues involved?
- What future work could come out of your project?

Know your references

Remember to read your references. You could legitimately be asked a question about any one of your cited references. Make sure you really do know what is in the key papers you have used. It is remarkable how often people cannot answer a question about a paper they have referenced many times in their thesis; possibly because they are genuinely too nervous to remember, possibly because they took the reference from a computer database without reading the paper.

Know your examiners

You will probably know in advance who your examiners are. If you do not it is worth finding out and learning as much as possible about them. Carry out a database search and find out what they have published and what

Fig. 10.1 Know your examiner.

their scientific interests are. This may give you some idea of their likely line of questioning, but do not be lulled into a false sense of security; it is dangerous to assume too much.

Formats of undergraduate and master's vivas vary, but most PhD or DPhil vivas are undertaken by two examiners: an 'internal' examiner from your university (but not department) and an 'external' examiner from outside your university. The external examiner is usually the person leading the viva.

Make sure your knowledge is up-to-date

Sod's law requires that any major breakthrough that occurs in your field and directly affects your results will happen on the morning of your viva, or possibly three or four days before. So make sure that your reading is up-to-date and you have a very good idea of what is happening currently in your chosen research area. You might not feel like it but, particularly in the week preceding your viva, go to the library and read any publications that might be relevant. This is an absolute must if you have had a long break between carrying out the research project and reaching the viva.

Taking Your Thesis or Dissertation to the Viva

Take a copy of your thesis or dissertation to the viva so you can refer to it when asked questions.

Taking Raw Data to the Viva

Occasionally examiners will request that a candidate bring along some of their raw data to the viva, such as photographs or traces. Even if you are not asked to, you might want to bring in raw data, your practical books or other material that relates to results your examiners may call into question. For example, if important details in a figure are difficult to make out because of problems with the original or with reproduction of it, you might want to bring the original so you can show the examiners that a detail such as a band or line is really present.

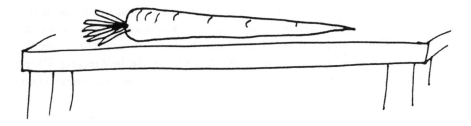

Fig. 10.2 Bring your raw data to the viva, if necessary.

The Day of the Viva

Although a little light reading will probably not hurt you, it is too late to cram in much more information the night before your viva. Relax and get yourself in the right frame of mind. Do not go out and get paralytically drunk with your friends. Do not get arrested. Do not have a row with your partner. Do not decide to murder your supervisor—what is done is done. Have a nice meal. Make sure you know exactly where the viva is and how long it will take you to get there. Make sure your clothes for the viva are ready to wear, are comfortable, and make you feel good. Make sure you have enough money to get to the viva, and for any phone calls you want to make after the viva. Make sure any paperwork or data you want to take with you are prepared. Also make sure you have a bottle of water, packet of tissues, throat lozenges, or whatever else you might need in the viva.

Relax. Remember that (1) many people have been in your position before and have survived, and (2) the examiners know perfectly well how you feel, and are generally sympathetic.

Finally the day of the viva comes. It is a good idea to arrive about twenty minutes early and to let the examiners know that you have done so. Take a few minutes to get your bearings, then head for the nearest lavatory. If you are due for a long viva you want to feel as comfortable as possible. Sitting cross-legged for five hours because you forgot to go to the loo beforehand is not feeling comfortable. Loos have mirrors, so you can check you look terrific and all your usual bodily parts are still there. You can also give yourself a big encouraging smile and check that your shoes are on the right way round and that you look respectable. It is difficult trying to answer a question about quantum mechanics when you suddenly notice you are wearing odd socks and have a toothpaste stain on your shirt.

A few minutes before the viva, wait where the person in charge of the exam can see you. You will be called into the examination room. Walk in. Sit down. The examiners will introduce themselves, and then, almost always, start with some fairly gentle questions to get the ball rolling. There is no need to burst into tears as you enter the room. The examiners know how you feel. Take a deep breath and off you go!

Fig. 10.3   Finally the day of the viva arrives.

If you feel nervous about your viva (and almost everyone does) count yourself lucky that you do not have to present a public defence of your work in front of family, friends, and the rugby team, as many students in the USA, Europe and Australasia have to.

In the Viva

Sound enthusiastic. We knew a student who was extremely bright and a very good scientist. She did not come across well in interviews or oral examinations because when she was nervous she appeared detached and uninterested. Interviewers felt she was bored by the topic and probably did not know much about it. In fact she had a passion for her subject and has gone on to a very successful academic career. Another student, probably not as able, did extremely well in oral examinations because he sounded enthusiastic about his project and so endeared himself to his examiners (who are only human, after all). You chose your subject because something about it appealed to you. Remember that enthusiasm and pass it on to the examiners. Enthusiasm is infectious. If you seem interested in your work they will be interested in both it and you.

If you are asked a question and have absolutely no idea of the answer, it is generally better to be honest rather than to try and bluff your way through. Bluffing is usually obvious and may get you into deeper water if the examiner decides to pursue your answer. If you disagree with an examiner, then by all means say so, providing you can justify your point of view.

After the Viva

For some exams you have to wait several days to find out your results. For others, such as PhDs, you will know the result the same day. For undergraduate degrees, such as BScs, BEngs, BAs, finding out your result will be the end of the story. For PhDs and a few MScs you will almost certainly have corrections to carry out—usually simply typographical errors. If you have submitted your work in a temporary binding you can simply make the corrections, print the new pages, and insert them before final binding of the thesis. If you have submitted a bound thesis or dissertation, the easiest way to make corrections is to print whatever text or figures are necessary and then glue them over the offending error. If you have to add extra material of a page or less simply stick it on the back of a page in the appropriate place in your thesis or dissertation. If more extensive corrections are required you may need to have your thesis re-bound to incorporate the changes. If you have to carry out further experimental work, liaise with your examiners about writing and re-binding your thesis. Your corrections, however trivial, must be checked and passed by an examiner before they can recommend the university to award you a degree.

Make your corrections immediately, because anything left for a long time becomes more of a chore and because universities often have strict rules on how much time is allowed for corrections. For example, the University of London requires minor PhD corrections to have been seen and checked by an examiner within four weeks of the viva.

Very rarely people feel their result was inappropriate. If you feel that your result is unfair take up the matter with your supervisor and university or departmental authorities immediately, who will look into it.

If You Should Fail

In the highly unlikely event that you fail your degree do not despair. Remember that you always have a variety of options and ways to deal with any situation, however depressing or hopeless it may seem. You need some time to think about what you want to do next. Other people have been in your position and have gone on to great success.

First find out your options, of which there are many. Talk to your supervisor or academic administrator to find out what academic routes are open to you. Discuss them with family and friends and other people who know you. University chaplains and counsellors are used to helping people with decisions and will be able to help you. Do not lose faith in yourself and remember that any problem always has more than one solution. Give yourself some time to look at different possibilities.

Common Mistakes

- Not reading the thesis beforehand
- Not knowing the references that have been cited in the thesis

- Not thinking about the background to a project

  Do not forget our advice about checking the mirror before you go into a viva. One of us was involved in examining a PhD student who was extremely nervous. He walked into the viva wearing a brilliant white shirt and black trousers—with his flies undone. The examiners were in agonies of indecision about whether or not to tell him. They decided not to as they thought it would make him more nervous than he already was. About 45 minutes into the viva the student noticed his flies were undone and spent the rest of his viva in an embarrassed haze trying to both answer the questions and decide whether or not to discreetly reach for the zip.

Key Points

- Prepare yourself for the viva: read your thesis
- Think about likely questions and ask your supervisor for helpful comments
- Think about the background to your research project
- Revise the key references that you have cited
- Ensure your reading is up-to-date
- Use your common sense and remember that lots of other people have survived the process

Chapter 11

SUPERVISION

Your supervisor is there to help and criticise your work on your thesis or dissertation. Before starting on your project, particularly if you are an undergraduate, you might only have seen your supervisor from a distance, probably giving a lecture during which you could sit passively and either listen (or not) to what they had to say. Your dealings with them as a supervisor are completely different. You are an active partner in a two-way relationship. If you contribute little or nothing, your supervisor will have little or nothing to supervise. The longer and more complicated your project, the more important your relationship with your supervisor is. Like any relationship, both sides have responsibilities and need to communicate effectively with each other.

Your Role

You are not expected to work in complete isolation and should receive support from your supervisor, but do not abuse this relationship by pestering them every five minutes with trivial problems that you could solve yourself. It is up to you to order your ideas and grapple with interpreting your experimental results.

Your supervisor will be responding to you and your work. If you come to the relationship with an enthusiasm for your work and a willingness to discuss it, your supervisor is more likely to take an interest in you and your project and therefore give better advice and criticism.

Fig. 11.1   Michael Bodkin (DPhil, pending) remembers he has a meeting with his supervisor.

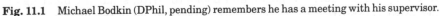

The Role of the Supervisor or Tutor

Your supervisor should provide helpful comments about how to sort the main ideas and themes of your work and structure your thesis. They should be prepared to read through drafts and the final version of your thesis or dissertation, and provide constructive criticism about the appropriateness of the contents and level of understanding you have shown, the thoroughness of your analysis, and the realism of your conclusions. They should also be able to advise you on your use of English and recommend suitable graphics and reference database programs to use.

It is not the supervisor's role to write the thesis for you. Your supervisor is there to guide your project, but they are not your personal secretary and proofreader.

See Your Supervisor Regularly With Drafts of Your Thesis or Dissertation

Make regular appointments with your supervisor to go through your work and make sure you keep them. Generally, you should expect to meet once a week, but this will depend on your individual circumstances. At the very least try and meet at least once every two weeks, whether you are studying for an undergraduate degree, master's or doctorate. Give your supervisor a copy of the draft you wish to discuss before you meet so they have time to read it. Apart from the constructive criticism a supervisor can provide, seeing them regularly will help to keep you motivated to produce your thesis or dissertation on time.

Make sure the draft you show your supervisor or tutor is spell-checked, well laid out and in a font that it is easy to read; use double line spacing so comments can be written easily on the draft. From a supervisor's point of view it is intensely irritating to have to spend time correcting spelling and formatting, which could easily have been done by the student. Supervisor's time is precious. In addition to looking after you they will have other students to supervise; research to carry out; teaching to do; talks to prepare; letters, grants, papers, reports, and books to write; scientific meetings to organise; committees to chair; papers to track down in the library; phones to answer; collaborators to e-mail; finance statements to work through; staff assessments to prepare; computers to mend; office supplies to fetch; works departments to interact with when their office door falls off its hinges; academic boards to sit on, etc., etc., etc. The list is endless. You are just one part of their responsibilities and you need them to focus on the contents of your thesis, not your spelling mistakes.

Problems With Supervisors

Some students feel, rightly or wrongly, that they and their work are being ignored by their supervisor—'*I sent my thesis draft to my supervisor ten months ago and he hasn't looked at it.*' If you feel that you are getting no feedback from your supervisor, arrange a meeting to discuss the problem. See if there is anything either of you can do to ease the situation; possibly you are being over anxious and a talk with your supervisor may put things into perspective. If, having talked to your supervisor, you still feel they are avoiding their responsibilities and not fulfilling their role, and they have made no indication that they will give more time to your work, then there are one or two things you can do to help get around the problem. Try meeting your supervisor and giving them your complete thesis plan, then set out a realistic timetable with them, detailing when you expect to have different chapters written. This will put your supervisor under gentle pressure. The timetable that the two of you have drawn up gives them deadlines by which they should have read each chapter and returned it with comments. At the same time, try to book a set of regular meetings in their diary for discussing your work. If your supervisor has a secretary, put these meetings into the secretary's diary as well—of course you must turn up to the meetings that you have arranged.

Most people survive their degree and their supervisor. In a very small number of cases the relationship with the supervisor breaks down. This might be the student's fault, or it might be the supervisor's. It is quite reasonable for a supervisor to get irritable if you mess them around with missed appointments and a sloppy attitude. If you are sure the problem lies with your supervisor rather than yourself the first thing to try doing is to talk with them about it. They may not know how you feel and need to be told there is a problem before they can do anything. Most supervisors want to do a good job but might, for personal reasons, find it difficult to do so. Remember that supervisors are as human as students. They make mistakes. They are emotional. They get divorced, and worry about the meaning of life.

You may feel loyal and sympathetic towards your supervisor, but this will not help you get your degree. If talking with your supervisor does not solve the problem you will have to go elsewhere for help. You could, by agreement with your supervisor, ask another member of the academic staff in your department to supervise you on an informal basis. If this is not possible you could see your director of studies, who should be able to sort out the problem either formally or informally. Sometimes students and supervisors suffer from serious personality clashes. By the time students are writing their thesis or dissertation this sort of problem has usually been resolved; if it has not, the student should contact their director of studies.

Obviously, if you are or have been suffering from sexual or racial harassment, or bullying of whatever sort, then you really do need to let someone know. If you feel difficult about going through formal channels, then talk to a sympathetic member of staff; this could an academic or it might be a chaplain or counsellor, whoever you feel most comfortable with.

Changing Supervisor

Occasionally students may have to formally change supervisor while writing their thesis or dissertation, for example, if their supervisor leaves the department suddenly. This is usually only an issue for PhD students who may take a few months to write up. It is the department's responsibility to allocate another supervisor. If a supervisor is leaving for another job, this has probably been on the cards for some time and the department should have already lined up a replacement. If the supervisor is suddenly taken ill or dies, the student might be left for a short time without guidance, but a new supervisor will be found as quickly as possible.

Key Points

- Remember how little time your supervisor has and streamline the process of reading your thesis drafts: spell-check and format your drafts and print them in double line spacing for easy reading; the supervisor needs to be able to concentrate on what you have written without being distracted by minor typographical errors
- If you feel your supervision is inadequate then first try talking to your supervisor. If this does not make a difference then you must get help from elsewhere; talk to a counsellor or your director of studies

A flatmate of one of the authors worked in a very successful and dynamic lab in Boston. The lab head had an unusual approach to management and care of his post-docs and graduate students. On one notable occasion when this flatmate had finished an arduous and demanding experiment, and it had not worked, the lab head rolled up a piece of X-ray film and walked round the lab trumpeting 'failure, failure'. If this happens to you, it's probably not unreasonable to feel aggrieved.

Chapter 12

RESOURCES

To work well you need to use all your resources efficiently; your most important resource is yourself.

Managing Your Time

Draw up a reasonable timetable for writing your thesis or dissertation. The amount of time you need for writing will vary according to the length of your project, ranging from a few weeks for a BSc or MSc project, to a few months for a doctoral study. You know at what time of the day you work best; make sure you keep this time clear for working. Pace yourself like an athlete in training. If you push yourself too hard you will burn out and go mad. If you do not push yourself at all you will not finish the dissertation or thesis on time, or perhaps ever. You need to set yourself a series of attainable goals for each work period you have scheduled. Decide what needs to be done, and plan to get it done in the time you have allowed yourself. You will of course often find you need more time to complete the task than you have planned, but you need a goal to aim for.

Managing Your Tasks

You need to be both systematic—making sure you do not overlook any details—and strategic—planning your work so you get it done in the most

Fig. 12.1   Manage your time sensibly.

efficient and therefore worry-free manner. There will be some things that are more pressing than others, so put these high on your list of priorities. Remember that you are not an automaton; you need stimulation and changes of activity, so if you have a particularly tedious or painstaking task to do it is a good idea to break it into sections and do a bit at a time in between other more interesting jobs. Preparing a figure can be a welcome break from typing or reading references.

The best way to crack a difficult point, or to work out exactly what you want to say, is not to sit at your keyboard for hours and hours trying to force an idea when you would rather go for a walk. Go for a walk. The mind is a strange and wonderful thing. It will often come up with the answer if you leave it to freewheel on its own for a while. If it has not come up with the answer by the end of your walk do not punish yourself—your mind will just jam up. Relax, have a drink, watch television, go to bed. The answer might well pop into your head in the morning.

Getting, Storing and Processing Information

You will need to know how and where to get information, to process this information efficiently, and to store it safely. Some of your information will be irreplaceable, some will be difficult and time-consuming to replace. Right from the very beginning of your project you need an organised regime for looking after your data and references.

Pens, paper and post-it notes

These are very basic resources, and you should have a ready supply of them. You need to be able to jot down bits of information and ideas whenever they come to you. Post-it notes are very useful for sticking notes in journals or books you are reading.

Notebooks for keeping ideas and for information

Keep an 'Ideas' notebook for putting down odd ideas about different aspects of your project, for example, points for the Introduction or Discussion. The more you learn and think about your project, the more you will want to discuss and explore different themes. You will have stray ideas—the first inklings of connections between different bits of information. These could occur to you at any time and you might have to scribble them down on the back of bus tickets and beer mats. If so, copy them into your Ideas notebook, or at least stick your bus tickets and beer mats into it as soon as possible so you do not lose them. Very often an idea that has been eluding you all day will come to you in the night. Scientists have their muses and should have something on which to write ideas when inspiration comes, so keep your Ideas book (with a pen or pencil) by your bed. If you wait until the morning the idea could well have vanished into thin air—or have become so deeply mixed up with that dream about you, Brad Pitt/Kylie Minogue/Mark Morrison/Jackie Chan and the Gold of Deadman's Bay, that you can no longer quite work out its relevance to your work on nuclear fusion.

You can use a different section of the same book for noting the main points from papers and articles you do not wish to photocopy, or for information that you come across in seminars or posters at scientific meetings.

Use a notebook which has the advantage of being bound so you are unlikely to lose pages from it, or use a ring-binder, which has the advantage of being flexible and easily organised and re-organised as you go along. If you are using a notebook, it can be useful to leave three or four pages blank at the beginning for a table of contents.

Keep your Ideas notebook and any other books well organised and safe, and if possible in a place where they will not be disturbed. Do not leave them lying around in pubs or on buses.

Making notes

Make your notes as easy to follow as possible. There are many different ways of taking notes; use whichever you find most comfortable. Use different coloured pens for different themes if it helps. Putting individual concepts or items of data into bullet points is a good way of summarising simple ideas. Flow diagrams help show sequences of events and connections between ideas. However you take notes, always put them in your own words and try not to write down too many details—if you are taking notes from a reference you can always go back and re-read the paper. Pick out the main points of interest, and summarise them. Never simply copy the ideas word for word, or you will be in danger of unconsciously plagiarising. It is very easy to copy your notes straight into your text forgetting that the words are somebody else's. Make sure you know exactly where your information came from—write down either the date and occasion, or the reference (see *Chapter 2: References*).

Some people like to use highlighting pens or underlining to pick out important points in photocopies of their references. Use this system if it is helpful, but do not highlight every detail and take care not to plagiarise when writing your text from these notes.

Textbooks, Dictionaries, English Usage Books, and Thesauruses

Keep to hand any regularly used textbooks or reference books. Buy a dictionary because you need to know the meaning of words you are using. As well as knowing what the words mean, you need to know how to use them, so it is a good idea to buy yourself an English usage book. A classic, which professional writers use, is Fowler's Modern English Usage (Oxford University Press). This is the recognised authority on conventions in current-day English, and is also an interesting reference book to browse through.

A thesaurus can be very useful if you find yourself using the same word over and over again and feel the need to vary your prose a bit. The word *thesaurus* comes from the Greek word meaning store, treasure, or storehouse. It is just that, a treasure house of words arranged into groups with similar meanings. If you are stuck for a word, or have a word that you know is not quite the right one, you can use a thesaurus to track down a suitable alternative.

Libraries

All libraries offer the same basic services, for example, providing textbooks and journals, ordering articles for you, allowing you to carry out reference database searches, and finding out about information available from other libraries. They also hold useful specialised reference books containing information such as citation abbreviations, SI unit abbreviations and definition lists, as well as general reference books such as encyclopaedias and dictionaries. One of the most useful resources you will find in the library is the librarian. They know how the library works and will be able to either find information for you, or show you how to find it for yourself. If you have only a vague idea that some information might exist, or just want to know what information might be useful to you, do not hesitate to ask the librarian. The more focused you are in your question, the easier they will find it to help you.

Carrying out a Literature Survey

A literature survey is simply a review of all the literature (books and journal articles, for example) that might be relevant to your project. Use the computers in your library, or any other suitable computer, to carry out a search of the literature databases (such as BIDS, Medline, Compendex, Inspec, or Chemical Abstracts) with key words from your subject area; also browse the World Wide Web using key words. It is up to you how many years back you go. If your key words have brought up huge numbers of references, then start by selecting only those most obviously relevant to your project, and the most recent. Be imaginative in your use of key words, and try searching with common abbreviations as well as complete words and phrases.

If you download reference details directly from a computer database into your own computer database, remove the odd bits of notation, such as hyphens, that you may not want. Read the references, not just the on-line abstracts.

Most of your references will come from books and journals or other sources of published information including other people's theses or dissertations, newspaper articles, conference proceedings and catalogues. You can also cite information from lectures, talks, or unpublished works if you have to, but it is best to stick to published material so the examiners can read your sources. If a work is not in the public domain, for example, if it is in a private collection, a private conversation, or a technical report that is only available on request, ask permission before citing it.

Building Your Own Library of Papers

During the course of your project you will come across papers in journals and other publications that you will want to use when you come to writing your dissertation or thesis, particularly for your Introduction and Discussion. It is a good idea to take photocopies of any papers you think will be useful to you. If you rely on their being available from your university or departmental library when you are writing up, you may find they have been taken out by someone else, or lost. Most scientists have had the experience of finding that the key paper they are after is in a set of journals that are 'out for binding' from the library. If you have your own copies, you will be sure of having them available when you need to use them.

As you collect your references, enter their details (for example, *Mason, N. and Hernandez, D. (1998). The Life Cycle of Landfish in East Coast Resort Towns. Landf. Tod. 6: 13-19*) into a computer file—preferably in a reference database program or, if not, at least a word processing file.

Copyright

Copyright protects the intellectual property of an author in whatever form it is recorded: books, journals, photographs, compilations, offline databases, computer programs, CD-ROMs, computer discs, etc. In Britain, copyright lasts for 70 years after the death of the author. Copyright means that no one can reproduce an author's work without having first gained permission. For private study, research, and criticism you are allowed to reproduce sections of someone else's work under a convention known as 'fair dealing' ('fair use' in the USA). Very briefly, fair dealing will allow you to photocopy other people's work for private study under the following guidelines:

Periodicals

Generally, the guidance is not to copy more than one article per issue.

Books and other non-periodical material

You can make one copy of a complete chapter in a book, or 5% of the book, whichever is the larger. For short books, reports, pamphlets, etc. the guidance is up to 10% if this is not more than 20 pages, or two pages if the original is very short.

Poems, essays, short stories, etc.

You can make one copy of any of these as long as they are no longer than 10 pages, in which case limit yourself to 5%.

Illustrations

It is acceptable to make one copy of a photograph or illustration for private study or research.

Ordinance survey maps

Do not make a copy larger than A4 size and no more than four copies.

Quotation

With most publications you can quote one extract of not more than 400 words. If you are quoting several extracts from the same piece, then not more than 300 words per extract. If quoting a poem, you can write up to 40 lines or the whole poem if it is less than 40 lines. You are free to quote as much as you like from HMSO publications and the Official Journal of the European Union.

Check with your library if you are at all unsure about copyright restrictions.

Information from Commercial Catalogues

Suppliers of equipment and materials, like chemicals or molecular biological resources, provide a lot of information in their catalogues, which could be useful when you are writing your Materials and Methods. Use the catalogues as another of your information sources, referencing if necessary of course.

Information from Microfiche

Until fairly recently some types of information, such as large bodies of data, were usually stored in a very small font on a type of transparent plastic card called a microfiche. To be able to see the microfiche you need to put it in a microfiche reader, which can also produce a printed copy for you to take away. Most large libraries either still have microfiche readers, or know where to find one. If you are using anything from microfiche, remember to reference it.

Information from CD-ROMs

An increasing number of important reference works are available on CD-ROM. You can access and use this information in much the same way you would that in a book. Find the CD-ROM you want and a computer that has a CD-ROM drive. Then work with the files on the CD-ROM in the same way that you would work with a floppy disk. Remember to reference anything you use from CD-ROM.

Computers

For some courses, not least of which is Computing, a computer is an essential everyday piece of equipment. For other courses, particularly the more theoretical ones, you might only ever use your computer as a magic typewriter. A surprising number of science students, especially biologists, have a phobia about computers. This is unfortunate as these are vital tools both for the research itself, and for writing reports—including dissertations and theses. We will not discuss uses of the computer specific to particular courses, but will confine ourselves to the use of the computer as a magic typewriter.

From the start, store as much of your information as possible on computer, as this will save you time when you come to write your dissertation or thesis at the end of your project. When you begin the writing proper you will be working most of the time on computer, word processing your information.

If you do not have your own computer, you will almost certainly have access to one in your department—know where it is and schedule in regular sessions so you can update your computer records and become familiar with how the machine works.

Guidelines for VDU Users

In the commercial world there are strict regulations governing how and for how long VDU operators should work. The regulations are there for a good reason. People are not machines; they need breaks, they need cups of coffee, they need to go to the toilet, fresh air, a change of scene from time to time.

Make sure your chair is comfortable. You need the lumbar region of your back to be well supported. The chair should be set to the right height, so that your thighs are horizontal and your lower legs and torso are vertical. The keyboard should be at such a height that your arms fall comfortably to your side, with your forearms resting horizontally. Keep you wrists as straight as possible and avoid having to twist them. Ensure there is enough space between you and the screen so that you do not have to squint to get it into focus.

The image on the screen should be stable with no flickering. Adjust the brightness and contrast so you can easily see what you are typing. Also make sure that the lighting in the room gives a good contrast between what is on the screen and the surrounding area. Most importantly, there should be no glare from electric or natural light reflecting from the screen, which will tire your eyes quickly.

Take regular breaks. It is best to take a number of shorter breaks frequently rather than waiting until you are tired and then taking a long break in order to recuperate. A break of about five or ten minutes every fifty or sixty minutes is recommended. Take these breaks away from the screen; either go for a walk, have a cup of tea, or perhaps do some work on your figures. You will be able to find more detailed information in the 'Health and Safety at Work' section in your library.

Thesis Templates

Some universities and colleges have what they call a 'template' for theses and dissertations, which will be available either on their home page (see

Fig. 12.2 Are you sitting comfortably?

our section on the World Wide Web, below) or on a floppy disk. Institutions provide templates to ensure their students' dissertations and theses conform to a standard layout. If using a template, no one has any excuse for missing out sections of their text or important details such as the information required on the title page. Find out if there is a template available for you, and if there is, use it.

Saving Your Information

Remember that computers crash quite regularly; whenever you stop to gather your thoughts for a second, or come to the end of a sentence or paragraph, give the save command. Get into a habit of saving regularly so that it becomes automatic. There is nothing more frustrating than stretching back with a satisfied yawn after three hours typing, thinking what a good evening's work you have done, to see it all disappear forever as your computer crashes and wipes all your un-saved work.

It is best to work on files that are on the hard disk and to keep the master copy of all your work there. Regularly copy from the hard disk to a floppy disk while you are working on your document. Copy your work onto at least three back-up disks. We have been told that most crashes of computer storage happen when copying from disks, so having just one back-up disk is not enough. If your one and only copy of your text is on a disk which becomes corrupted, you will have lost everything. Keep your back-up disks safely stored, each one in different locations in case one is lost or stolen.

If, when you insert a floppy disk into the computer, you are told the disk is unreadable, do not panic. If you just slide back the metal shutter that protects the disk you may be able to gently blow away dust that is causing the problem. If this does not help there are many programs around that can retrieve the contents of damaged disks—ask your computer department to help you. Do not reinitialise the disk, as your computer may suggest, or you will wipe all the information from it.

Some people recommend that you keep graphics files on separate disks from your text files, so that if one of the disks corrupts, at least you will not have lost both text and graphics.

Keeping Printed Copies of Your Work

Do not rely totally on your computer's hard disk and your back-up disks. You might never have any trouble with either of them, but if you do, and you do not have any other record of your work, you will have to start from scratch again. Print a 'hard copy' of your text at regular intervals; 'hard copy' is just computer-speak for what most people would call a printed-on-paper. If the worst-case scenario happens and all your back-up disks corrupt and your computer melts because of a fire in the Chemistry Department, then it is likely that someone at your college or university will have a scanner (see below), and you can at least scan the printed pages of your draft into another computer file. This will save you having to re-type everything.

Filing Your Work in Folders or Directories

Organise a sensible archiving system for your computer files. This will help you see what files you have and which draft you are working on. It will also help you to avoid losing valuable files, which you might delete by mistake.

The 'containers' for individual computer files are called 'folders' on Apple Macintosh computers, and 'directories' on PCs. Organise your files into different folders or directories, and name the folders or directories so that their contents are obvious.

As well as naming your files (and folders or directories), get into the habit of dating everything, either within the text itself (you could put this information in the header or footer zones) or in the file name. Remember to change the date when you create a new version so you do not risk muddling multiple versions of the same file. It is a good idea to have a 'current' folder or directory for your most recent work, and 'old' folder or directory for older drafts, which you still want to hang on to. Do not be afraid to get rid of information once it is no longer any use to you, so clear out your 'old' folder or directory periodically, or it will just get cluttered with redundant information.

Typing

We are assuming you will be typing your dissertation or thesis yourself. How well you can type will affect the speed and ease with which you produce your text. Two-fingered typing will get you through, but the more fingers you can use and the more accurately you can use them the easier the task will be. It will save you time in the long run if you learn how to type to a reasonable standard before starting to write your text. You could go on a typing course, or use one of the typing tutor programs available for use with most computers. You could ask the computing department in your institution to recommend a good one.

Word Processing

Fortunately for the computer-phobic, word processing programs are produced for everyone to use, from sous chefs to sky-divers, so they are easy and accessible for beginners. We have assumed that you have some basic knowledge of word processing. You probably know that word processing programs allow you to change what you have written easily, rather than having to type the whole thing out again, and that you can move chunks of your text around from place to place with the 'cut' and 'paste' commands, and generally play around with your text in a way that people could only have dreamed of a few years ago.

How much you learn about your word processing program is up to you, but learning a few commands, other than simply how to type and cut and paste, will save you a lot of time when you write your dissertation or thesis and will help you produce a much more professional looking manuscript.

All word processing programs are slightly different, so we will just give a few of the common functions rather than going into the details of the commands. The best advice is to read the instruction manual that comes with your word processing program, to use the 'help' option in the program, and to play around with the program experimenting with the different commands. Learn how to find any specialised symbols you need, for example, mathematical symbols, Greek letters, and accents. Learn about commands using combinations of keys rather than working with the mouse alone, as key commands are often faster, and make sure you know how to use tools such as the spell-checker and commands for layout (see *Chapter 13: Layout*). Here are a few of the keys and commands with which you should be familiar.

The SHIFT and CAPS LOCK keys

The SHIFT key makes the letters and numbers on your keyboard type in upper case—that is the letters are typed as capitals—and the numbers are typed as a range of different symbols. The CAPS LOCK key 'locks' all the letters on your keyboard so they are typed as capitals, but this command will only work with letters, and will not alter numbers. Remove this command by pressing the CAPS LOCK key a second time.

The TAB key

The TAB key allows you to indent, or move in, your text to a particular point. Most word processing programs are set up to have automatic Tabs that make the cursor stop at the same set of points on every line, so your paragraphs, for example, can always start at the same point. The Tabs in your document are all moveable and removable. You can put them where you want and change their settings. Look at your manual or the on-line help to learn more about the Tab key. It is very useful when writing text and producing tables.

Finding and replacing

If you decide you need to change a particular word or phrase you have typed many times, for example, if you want to change every *didn't* into a *did not,* your word processor will have a command (for example, 'FIND AND REPLACE') that allows the computer to automatically find the first word or phrase and replace it with the second. If you only want to replace full words, set the program appropriately: it is very frustrating to find you have not only changed, for example, *end* to *beginning* throughout your text, but also *dependant,* to *depbeginningant,* and *blend,* to *blbeginning.* You can get around this either by making sure the program is set to only replace full words, or by putting a space either side of the word to be replaced.

Headers and footers

The headers and footers are the spaces at the top and the bottom of the page (both on screen and on paper). You probably will not use these for writing a thesis or dissertation, but some people find them useful for holding the date or version number of the draft they are working on. Your word processor will have simple commands to allow you to use them.

Giving one simple command to write something long and complicated

If you have a long complicated word or phrase you use frequently, it is a nuisance to have to type it in full every time you need it. You can save yourself time and finger-work by making use of commands on your word processor known as macros, or short cuts, which enable you to have the computer add a particular word or phrase to the text every time you give a simple command. For example, you can arrange things so that every time you need to write *Saccharomyces cerevisiae, scanning tunnelling electronmicroscopy,* or *electromagnetic resonance imaging,* you just press the control key followed by a key of your choice. Have a look at the 'help' section of your word processing program.

Using a spell-checker

All word processing programs have spell-checkers and so it is inexcusable to make spelling mistakes in a thesis or dissertation. When you are setting your spell-checker be aware of the differences between American and British English, most of which are to do with spelling. Set your spell-checker appropriately: if you are studying in Britain it is best to use British English.

For scientists, one of the most important commands to learn is that which lets you create what is called a custom dictionary. This is your private dictionary for the spell-checker program and you can include all your favourite words which are unlikely to be in the computer's own dictionary. If you spell-check with your custom dictionary you can avoid stopping at every second or third word because the program does not recognise scientific words like *Fischer-Tropsch process, Lotka-Volterra mechanism, asymptotic series,* or *zwitterion.*

Using a thesaurus

Many word processing programs have a thesaurus. If yours has one make use of it when you find yourself using the same word over and over again.

Using a grammar-checker

It can be interesting and instructive to use a grammar-checker on odd bits of text, but it is not worth using one to check your whole text; it will take you too long and will usually only tell you there is a problem, giving you no idea as to a solution. Grammar-checkers pick up only some, not all, of the likely mistakes, for example:

> While undergoing treatment, the doctors found that their patients'
> recovery rate was improved by listening to classical music...

This is grammatically correct but is probably not what the writer
would wish to say. It means that the doctors were undergoing treatment
when they made their finding (see the section on dangling participles in
Chapter 14: The Use of English).

Number headings automatically—outliners

Most word processors have a command, often called 'outliner', that
automatically numbers your headings:

**THE CHANGING DEMOGRAPHICS OF URBAN
LANDFISH (*THESIS TITLE*)**

1.   **Problems facing urban landfish populations (*first
     chapter heading*)**
1.1 **Predation by foxes (*first sub-division*)**
1.2 **Dioxin contamination of urban habitats (*second sub-
     division*)**
1.3 **Psychological factors (*third sub-division*)**
     1.3.1 Fast food outlets (*first sub-sub-division*)
     1.3.2 Alcohol abuse (*second sub-sub-division*)
     1.3.3 Low status in society (*third sub-sub-division*)

If you move the headings around they will automatically be re-numbered.
This can be useful if you have your notes on computer and want to rearrange them as your planning progresses. These commands are different
for each program so read the manual or on-line help, or ask your computing
department for assistance.

Number headings automatically to create a Table of Contents

Many programs have an extremely useful command that not only numbers
your headings, but remembers them, notes which page number the headings
are on, and automatically creates a Table of Contents for you. This command

can save you an enormous amount of time when you come to finish your thesis, and is worth finding out about right now.

Word processing with mathematical symbols

All word processors allow you to put in simple mathematical symbols and write out equations, but most programs have been designed for using everyday English, and so typing the symbols using a normal word processing program can take what seems like forever. You will probably have to change to a special font and spend ages getting the symbols the correct size. Many scientists therefore work with word processing programs that have been specifically designed for easily and speedily incorporating mathematical symbols. If you are in any doubt about which program to use, talk to your supervisor or ask your computer department for help. We strongly recommend that you use one of these programs if you are writing a project with a large mathematical component.

Reference Database Programs

In the old pre-computerised days of thesis writing, every time you cited a paper in the text you may have had to write out the citation in full and then type the full reference details at the end of your thesis. Typing long citations many times (for example, *Gerwürtztraminer and Cabernet-Sauvignon, 1998*) is a nuisance and can lead to typing mistakes. It is also easy to forget to add or remove references from your reference list when you edit citations in your text. Reference database programs overcome these problems.

With a reference database program, all you have to do is type in the full reference details once. The reference is put into a 'library' on the computer and is given its own unique tag. When you cite the reference in the text, you just copy and paste the tag (which can include the name, date, and a symbol)—no more typing out names or dates or numbers. When you come to print the thesis, the program automatically puts the citations and full reference details into any format you like, only giving details of the citations you have placed in the text.

A further advantage of using a reference database program is that most of them allow you to automatically import reference details from the commonly used reference databases, such as BIDS or MEDLINE. This reduces the chance of typographical errors occurring and cuts down on the amount of typing you have to do. However, remember to take out any bits of notation, such as hyphens, which are often imported along with the reference details.

When you have formatted your references and are ready to print a draft, first check that all the references have in fact been properly formatted. If they have not been, the tag will not have been replaced by the appropriate citation. Run a 'find' for the tags symbols to detect any stray ones.

Most reference database programs are very easy to use and are invaluable to anyone citing a lot of references. Learn to use a reference database program as soon as possible. If you go on to further study you can build up a large library of such references over the years.

Spreadsheet Programs

These are wonderful programs that are ideal for storing large amounts of data suitable for placing in a table. As well as giving you a neat and tidy way of presenting lots of numbers or other information, spreadsheets also allow you to carry out mathematical processes automatically: for example, you could ask the program to multiply all the figures in column 1 by 537, and to enter the results into column 3. The program will do this, and so remove any likely errors you might make doing the same thing with a calculator and entering in the results by hand.

You can insert your data into a spreadsheet, while you are gathering them, and then import the spreadsheet directly into your text as a table when you come to write your thesis. This will save you time, and cuts down on the chance of making a mistake while re-typing the data.

A number of spreadsheet programs exist, so we are not giving specific instructions about any of them. Most of them are not difficult to use for simple applications and can provide a professional-looking way of presenting a large amount of data.

Graph Programs

There are many graph programs around that will automatically create graphs from data that you enter. These programs can be simple to use, at least for basic tasks, and if used well will produce beautifully drawn curves and lines, saving you a lot of time with graph paper and rulers. Usually, the programs work by having a spreadsheet into which you place your data, then you have a choice of graphs that the computer can create from these data. For example, you could produce a histogram, or a pie chart, or a scatter graph, or a curve using points either with or without error bars. Usually your data are only appropriate for one type of graph, so the decision is easy, but if in doubt go to your supervisor for guidance.

Remember to fully annotate your graph when using these programs, for example, say what the x and y axes represent and clearly indicate a selection of values along these axes.

Be wary of going over the top using the facilities that are available to you on these programs. Seminar speakers quite often make the mistake of presenting amazing multicoloured, unnecessarily three-dimensional graphs, in which the actual data are lost in a mass of fancy graphics. The simplest graphs are usually the most striking and effective way of presenting your data (see *Chapter 7: Figures and Tables*).

Graphics and Drawing Programs

Like all other computer programs, graphics and drawing programs come in various shades of complexity. Small children play with simple graphics programs, producing fat ellipses and large squares and telling people it is a picture of Mummy and Daddy. At the other end of the expertise scale, professional graphics and video design artists produce very complex graphics with similar programs. Most of us lie somewhere in between these two extremes.

We are going to generalise wildly about these programs and really you need to find out for yourself what your own resources are: graphics programs allow you to produce complex images from relatively simple components such as circles and squares and straight lines, with or without colour. They are most useful for producing line drawings and diagrams, and for simple annotation of previously scanned images. Drawing programs tend to be used more by professional artists and are often a lot more flexible but less user-friendly. Many scanners use drawing programs for altering and manipulating scanned images, for example, to reduce the intensity of the background or increase the intensity of the foreground.

The trick to working with all graphics and drawing programs is to get hold of the manual or read the on-line help, and, most importantly, play with the programs. We encourage our students to design their party invitations, play around making Christmas cards, send fancy letter headings to their friends using the graphics and drawing programs that we have, because this is an excellent way to become familiar and at ease with the process (see *Chapter 7: Figures and Tables*).

Scanners

A scanner is a machine that will scan or copy text or an image into a file on your computer. Scan text, for example, if you need a long passage of typed text but do not want to type it into a computer file yourself. Scan images into a computer file, for example, data from a photograph, or X-ray film. If you need to annotate the image, perhaps adding arrows to highlight particular features, or by writing on numbers or text, you can then use a drawing or graphics program.

Computer Networks

Many computers in university and college departments are linked to a network of other computers. The other computers may contain useful databases or can simply act as addresses for receiving and sending e-mail.

The Internet

The Internet is a very large computer network. It is a global communication network that connects computers, and computer users, throughout the world. The Internet is a wonderful, user-friendly creation and you should learn the basics of sending and receiving messages (such as e-mail) as soon as possible. It is an excellent way to stay in touch with friends in distant places and in the next office.

The World Wide Web

The Internet started in the late 1960s. More recently came the advent of the World Wide Web. This uses the Internet but instead of simply allowing us to send text and computing files around the world. It provides the person looking at the computer screen with a graphical display (a picture) related to the database or computer site that they are communicating with. This is the 'home page' for that particular address on the World Wide Web. The great advantage of these 'home pages' is that you can very easily extract data from them, or use the programs that they contain—usually just by moving the cursor over the item you are interested in and clicking.

Using the World Wide Web is fun and easy

Once you have found a computer that is networked to the World Wide Web you need to open the program that allows you to see the pictures of the home pages. Various programs, all very easy to use, allow you to do this. Ask someone who uses the Web to show you how to open the program. As the program starts running, the first thing you will probably see is the home page for your university or department, which normally shows a picture with information such as the name of the institute, what you can study there, the names of the staff members. If you look at the text you will see that some of it is underlined and probably in a different colour from the surrounding text. When you put the cursor on top of this text, and click the mouse, you will go onto the next 'page' that provides information relevant to this text. You can click onto the 'back' arrow at the top of the page to go back or 'forward' to go forward to pages you have visited. The 'home' arrow takes you back to your home page. In this way you are navigating around or 'surfing' the Web... Cool.

You can find information about anything *on the Web*

There are millions of home pages out there. Some are set up by institutions, such as large American universities; others by individuals, such as small Japanese school boys. Each home page has a unique address; if you look at the top of your home page you will see its address and it will look something like:

  http://www.ic.ac.uk/

which is the address of the home page for Imperial College, London.

If you know the address of an interesting home page, you can type it into the address space at the top of the page and the computer will take you straight there. If it is a place you regularly visit, ask your local computer person how to create a 'bookmark' so that this address is stored in a type of computer address book, and you do not have to keep typing it each time you use it.

The World Wide Web is full of home pages containing interesting information. You will see an option called 'search' (or something similar—ask your computer person for help). Open this option and type in a key word, or several words on a subject of interest. Click on the 'search' button and the computer will give you a list of sites that contain home pages with relevant information.

Each country has different laws about exactly what you can download from the Web; in the UK downloading hard-core pornography, for example, is illegal.

Reference the name and address of home pages that you have used

While carrying out your research project or writing your thesis, you may have used either information or programs from different web sites. You need to reference these sites by giving their name and address (http://www....) so that any examiner or fellow scientist can see exactly what you did. You could put all the addresses into a table in your Materials and Methods, or in an Appendix or the Reference list or in footnotes at the bottom of the page.

Learn to Love Your Computer

However new your computer is, it will crash from time to time and your programs will develop glitches. You need to learn at least the basics of how your computer and programs work so that you can deal with these problems when they come up. We are not advising that you become a computer engineer on top of all your other studies, but if you make the effort to familiarise yourself with your equipment and learn to recognise easily fixed problems, you will save a lot of time that would otherwise be spent waiting for someone else to sort out the computer. Most of the time the trouble is fairly minor. Get hold of a manual for your computer and the programs you use regularly. Have a look through the manuals so you get the gist of the contents, and keep them somewhere where you can get at them quickly if you have a problem. Most manuals have sections entitled 'Troubleshooting' (or something similar) that can help you fix the common glitches. If your computer develops a major glitch then you have no choice but to call in an expert as soon as possible.

Common Mistakes

- Running out of time for writing up before a deadline
- Working in uncomfortable conditions: spending long hours in a bad posture in front of the computer will lead to backache and eye strain
  Not all computer problems are soluble. A former colleague who is now an important professor working in the USA did his PhD in England in the 1970s in a department that was one of the first to get a large mainframe computer. Working late in the lab one night, he accidentally knocked a litre of boric acid down into the back of the computer. Afraid to admit to the disaster he watched, terrified, as this expensive new computer developed more and more faults over the next few days. He developed bad eczema as his situation—and that of the computer—worsened. Finally, someone took the back off the computer and discovered the rotting heap of wires inside. The graduate student owned up, and his eczema improved, but not his supervisor's temper.
- If using a reference database program, do not leave any strange notation such as hyphens or hatch signs in the final formatted version of your text

Key Points

- Set a realistic timetable for achieving realistic goals
- Notebooks are an invaluable and often irreplaceable source of information, so make an effort to keep them safe, and write some sort of contact details in the book in case it gets lost
  As an undergraduate one of us had a bag containing an entire year's worth of notes stolen from a car a week before his exams. It was not a happy experience, particularly as all the notes had been borrowed from a friend.
- Make full use of your library
- Put your reference details into a computer file immediately
- Take care NEVER to plagiarise when copying from notes or references
- Use a thesis template if it is available for you
- Save all computer files on a hard disk and at least three floppy disks, and keep printed copies of your work
- Learn to use your word processing program beyond the basic commands, and make full use of other programs such as reference database, graphics and graph programs

Chapter 13

LAYOUT

A manuscript that is clearly laid out will be far easier to read than one thrown down onto the paper in any which way. Good presentation shows that you have taken care over every aspect of your project: if your manuscript is laid out logically and consistently it will help both you and your examiners to take your work seriously. It will also be easier for you to spot any omissions or mistakes when you are checking your drafts. When you are planning and writing your dissertation or thesis, the quality of your layout is nearly as important as the organisation of your ideas.

Good layout needs good planning. Your department or university may have particular rules covering the layout of your text. Find out if there are any in your case, and if there are act upon them. Decide on the overall format of your document, the size of margins, headers, and footers, line spacing, etc. before you start writing. Changing the basic layout of your text halfway through or at the end of your writing is very time-consuming. We strongly recommend having a look at previous dissertations or theses in your subject to see what you do and do not like about their presentation.

You should be aiming to produce a document that is stylistically consistent throughout, looks well organised, and is pleasing to read. We looked at some basic word processing commands in the previous chapter, here we consider basic formatting.

Fonts and Line Spacing

Your choice of font and line spacing will affect the number of pages in your final manuscript. This does not matter as long as you have kept to your word limit if you have one (word limits are usually set for undergraduate and some MSc theses or dissertations, but not DPhil and PhD theses).

Fonts

'Font' basically just means typeface. Each font has a name to identify it, for example:

> This is Times font
> This is Geneva font
> This is New York font
> This is Avant Garde font

Your word processing program will have a number of fonts available in different sizes. The size of your characters is measured in either points or pitch. Very roughly, a single point is equivalent to 0.35 mm. 'Pitch' measures the size of the characters in characters per inch and is really only relevant to typewriters and older daisywheel printers, which print each letter with the same spacing.

Choose a font that looks clear and is easy to read when it is printed (we do not recommend fonts like 𝔒𝔩𝔡 𝔈𝔫𝔤𝔩𝔦𝔰𝔥 𝔗𝔢𝔵𝔱). Some fonts look good on the screen, but are not so pleasing on the printed page, and vice versa, for example, Geneva looks much better than Helvetica on some computers screens, but not as readable when printed on some printers. When choosing a font, type a phrase or two to see how it looks both on screen and on paper. Try 'The quick brown fox jumps over the lazy dog.', which contains all the letters of the alphabet. Here are a few fonts in different point sizes:

> The quick brown fox jumps over the lazy dog. (Times, 10 point)
> The quick brown fox jumps over the lazy dog. (Times, 12 point)
> The quick brown fox jumps over the lazy dog. (Helvetica, 10 point)
> The quick brown fox jumps over the lazy dog. (Helvetica, 12 point)

```
The  quick  brown  fox  jumps  over  the  lazy  dog.
(Courier, 10 point)
```
The quick brown fox jumps over the lazy dog.
(Courier, 12 point)

The quick brown fox jumps over the lazy dog.
(New York, 10 point)

The quick brown fox jumps over the lazy dog.
(New York, 12 point)

**The quick brown fox jumps over the lazy dog.
(Chicago, 10 point)**

**The quick brown fox jumps over the lazy dog.
(Chicago, 12 point)**

Some fonts space the characters according to their individual size, so, for example, an 'm' is given more space than an 'l'. Other fonts space characters so that each one takes up the same sized space, like on a typewriter. Generally, spacing according to the size of the character is much more pleasing to read than using one space per character.

This is printed using the font Times which spaces characters according to their size.
```
This is printed using the font Courier which gives
each character the same amount of space.
```

Fonts such as 12 point Times or 10 point Helvetica, or 12 point Helvetica, like this, all look pretty good. If you make the font too small, it will be very difficult to read (7 point Times). If you make the font too large, it will look as if you are writing a children's book rather than a scientific document (18 point Times). Once you have chosen a font and font size stick to it, altering it will throw out the formatting of your text and you will have to spend ages re-formatting.

Line spacing

Word processors use three standard options for line spacing:

Single,

> We hybridised the 984-bp mouse *DJHD* ORF probe to a somatic cell hybrid panel containing all the mouse chromosomes on a human background (Table 2.3). The probe detected five *BamH* I fragments of 12, 10, 8, 7, 4 and 3 kb in the total mouse track and the same five fragments in the cell line containing mouse chromosome 12 only on a human background. These five fragments were not present in digests of various human DNAs. These data indicate *DJHD* is present on mouse chromosome 12.

one-and-a-half,

> We hybridised the 984-bp mouse *DJHD* ORF probe to a somatic cell hybrid panel containing all the mouse chromosomes on a human background (Table 2.3). The probe detected five *BamH* I fragments of 12, 10, 8, 7, 4 and 3 kb in the total mouse track and the same five fragments in the cell line containing mouse chromosome 12 only on a human background. These five fragments were not present in digests of various human DNAs. These data indicate *DJHD* is present on mouse chromosome 12.

and double,

> We hybridised the 984-bp mouse *DJHD* ORF probe to a somatic cell hybrid panel containing all the mouse chromosomes on a human background (Table 2.3). The probe detected five *BamH* I fragments of 12, 10, 8, 7, 4 and 3 kb in the total mouse track and the same five fragments in the cell line containing mouse chromosome 12 only on a human background. These five fragments were not present in digests of various human DNAs. These data indicate *DJHD* is present on mouse chromosome 12.

Your department or university will probably have a rule about the line spacing of your manuscript; usually double or one-and-a-half line spacing are specified as they are easier to read (these are also easier to edit when writing corrections on drafts).

Margins, Headers, and Footers

You text should look comfortably settled on the page, so you need reasonable sized margins, headers (the space at the top of the page) and footers (the space at the bottom of the page). About 3 cm is a reasonable width all round except that you will need a slightly larger margin on the left of your page so you will not lose any text in the binding; set a left margin of at least 4 cm. Most universities and colleges have rules about margin sizes. Find out if there are any in your case, and make sure you know what they are.

You can put text in the Header and Footer. Some people type the chapter title or number in a small font in the header or footer zone, so the reader can tell roughly where they are when flipping through their manuscript.

Alignment of Text

Aligning to margins

There are different ways you can align, or 'justify', your text. With the majority of your text you will align it to the left margin:

2.3.1 Veronica-Fisher extrapolation
By this method the principal error functions of a given (low order) numerical method are approximated, thereby accelerating its convergence. Suitable linear combinations of approximate solutions obtained by using the low order method on different (uniform) meshes are calculated to complete this process. The Veronica-Fisher extrapolation depends on the knowledge of the powers of j appearing in the error expression for the low order method.

or possibly so that all lines start at the left margin and finish at right margin:

2.3.1 Veronica-Fisher extrapolation
By this method the principal error functions of a given (low order) numerical method are approximated, thereby accelerating its convergence. Suitable linear combinations of approximate solutions obtained by using the low order method on different (uniform) meshes are calculated to complete this process. The Veronica-Fisher extrapolation depends on the knowledge of the powers of j appearing in the error expression for the low order method.

You might want to align a heading to the centre of your text. If you do this, only align the heading, and not the text underneath it as we have done below:

2.3.1 Veronica-Fisher extrapolation
By this method the principal error functions of a given (low order) numerical method are approximated, thereby accelerating its convergence. Suitable linear combinations of approximate solutions obtained by using the low order method on different (uniform) meshes are calculated to complete this process. The Veronica-Fisher extrapolation depends on the knowledge of the powers of j appearing in the error expression for the low order method.

It is also possible to align to the right margin, although this would look odd for your main text.

2.3.1 Veronica-Fisher extrapolation
By this method the principal error functions of a given (low order) numerical method are approximated, thereby accelerating its convergence. Suitable linear combinations of approximate solutions obtained by using the low order method on different (uniform) meshes are calculated to complete this process. The Veronica-Fisher extrapolation depends on the

knowledge of the powers of *j* appearing in the error expression
for the low order method.

Indenting

Indenting just means moving a section of your text in from the margin a
little. The easiest way to indent is to use the automatic tab settings on
your word processor or to set a tab at the point to which you want to indent
(see *Chapter 12: Resources*). When you are typing all you have to do is hit
the tab key and the cursor will move to the tab point. The tab only indents
one line. If you need to indent several lines of text, move your margins at
these sections rather than having to tab each line that you type.

Indenting at the beginning of paragraphs

It is normal to indicate the beginning of a new paragraph by indenting it
about five character spaces. Use a tab key at the start of the paragraph to
create the indent.

>This paragraph, for example, is indented and is distinct from
>the previous paragraph.

Alternatively you need not indent, but can leave a line between paragraphs,
but this may make your text rather long.

Indenting quotes and examples

To distinguish quotes and examples from new paragraphs it is conventional
to indent the left margin of the complete quote by about ten character
spaces, for example:

>This is the sort of indentation you might use when giving a quote.
>The whole paragraph or section is indented more than the rest of
>the text and has been indented by changing the margins here.

Alternatively,

> For this paragraph, we have moved the margins in to indicate that we are giving a quote, and we have also used the tab key on the first line to indicate that it is a new paragraph.
>
> This is useful if you are writing a long quote made up of many paragraphs or sections, because you need to show that they are all separate.

Using different margins and indents

If you play around with your indent and margin commands (read the on-line help or manual of your word processing program) you will find commands that allow you to indent the top line a different distance from the rest of the paragraph. This is particularly useful for formatting lists of references where you want to ensure each entry is clearly visible and separate, but you do not want to waste space by putting extra lines between the entries, for example:

Kusumi, K., Pratt, S.C. and McCartly, M.I. (1999). A general high-efficiency procedure for production of landfish cell hybrids. *Landf. Tod.* 5: 119–125.

Hill, A., Hooton, F., Knight, A. and Collinge, J. (1997). A panel of somatic cell hybrids containing landfish chromosome 12. *Landf. Genet.* 115: 15–31.

Mahal, S., Asante, E. and Ashworth, T. (1996). Landfish gene transfer by means of cell fusion. *J. Landf. Sci.* 4: 8–12.

Strivens, M., Peters, J., Ball, S., Booker, D., Morse, S. and Brown, S. (1998). Two dominant-acting selectable markers for gene transfer in landfish cells. In *A handbook of laboratory studies*. (MUT Press Oxford).

Here the first line starts at the left margin, and all subsequent lines are indented approximately five spaces.

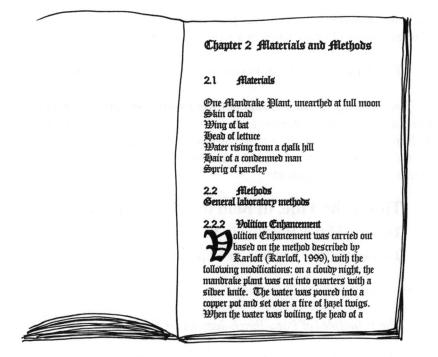

Fig. 13.1   Use a font and layout that are clear and easy to read.

Titles and Headings

Format

There are no rules about how you capitalise the words in your heading, but make sure your headings are consistent. Decide on a format and stick to it: use the same font and keep all headings in the same tense and otherwise stylistically similar; it is not normal to put a full stop at the end of your heading or title. For headings and major titles it is conventional to capitalise the first letter of words except prepositions and conjunctions (unless they are the first word of the title):

The Principles of Para-Psychology in an Urban Setting

Although many word processing programs have commands that capitalise the first letter of every word in a title or heading:

The Principles Of Para-Psychology In An Urban Setting

Your headings need to stand out from your text, but do not go over the top playing around with your fonts, otherwise your titles and headings will look messy and out of place with the rest of the text. The style we favour is for titles to be the same font, in bold, and either the same size or a couple of points larger than the text:

This is the Title, in Bold in 14 point Times
This is a Major Heading in 12 point Times
This is a minor heading in 12 point Times

Alternatively, you could use italics:

The Principles of Para-Psychology in an Urban Setting

Underlining looks fussy and we do not advise using it:

The Principles of Para-Psychology in an Urban Setting

We do not recommend the use of capitals because they look as if you are shouting at the reader, and no one likes being shouted at:

THE PRINCIPLES OF PARA-PSYCHOLOGY IN AN URBAN SETTING

Shadowed looks like a party invitation rather than a heading:

The Principles of Para-Psychology in an Urban Setting

Having said all this, you can use any combination that you think looks good. We do not think this one does:

The Principles of Para-Psychology in an Urban Setting

If a number forms part of your title it is best to write it out: *One* rather than *1,* and *ninety-five* rather than *95.*

An Experiment on Five Undergraduates Involving Electricity

Whichever style you use, bold, italics or the style of your choice, do not mix styles in different headings as this looks very messy.

Numbering: the title, headings and sub-headings

Do not number your title. In science theses or dissertation it is normal to number the chapter headings.

When you divide your chapters into sections, make use of sub-headings. To make it easier to follow the division of the headings and to help create the Table of Contents, sub-headings are usually numbered. The easiest and most logical way of numbering is by the chapter number, followed by the number of the sub-heading, for example, the first sub-heading of Chapter 1 would be 1.1, the next sub-heading as 1.2, etc. Within these sub-headings you can further sub-divide to 1.1.1 and 1.1.2, etc. as necessary. Do not put a full stop after the number. In the following example, the different category of heading are indicated in italics:

The Development of the Cheam Moog Genesis Technique (*thesis title*)

1   Problems with traditional Moog Genesis (*first chapter heading*)

1.1 The Woodward paradox (*first division*)

1.2 Instability in parallel M fields (*second division*)

1.3 Intermittent Rabbs emissions (*third division*)

    1.3.1 Edwards modulation (*first sub-division*)

    1.3.2 Batcock wobble (*second sub-division*)

    1.3.3 Linked field modification (*third sub-division*)

It is best not go beyond two sub-divisions (i.e. not beyond 1.1.1.1 or 5.4.6.3 to 1.1.1.1.1 or 5.4.6.3.7). You do not need to provide a numbered sub-heading for every tiny point you are making. If you really feel that you absolutely have to put in another sub-division and therefore another heading, then do not number the heading.

Separating Sections and Chapters

New chapters should start on a new page. Each new section should have a heading and be clearly distinct from the previous section, although it does not need to start on a new page. Never have a heading on the last line of a page.

Inserting Page Numbers

One of us, we will not say who, did not bother to find out how to automatically insert page numbers into their text when they first started using a word processor. He had to spend hours writing them in by hand after the text had been printed. Find out how to insert page numbers. If you do not put page numbers in and the pages become mixed up after being dropped on the floor, it is a nightmare trying to get them back in the right order. You can put page numbers at either the top or the bottom of the page in your header or footer zone, there are no rules, but make sure the number is easy to see; avoid the left side of the page, close to the binding.

You can leave the first page blank if you prefer and number from page one of the Introduction, or from the opening sections. Do not number your title page.

Highlighting Text

Occasionally you may need to highlight a word or phrase within your text. Using a **bold font** is usually the neatest way of doing this, but you could use *italics*. <u>Underlining</u> tends to look messy. Do not highlight words too often or they lose their impact and your text looks cluttered.

Bullet Points

Not all word processing programs have these, but play around and see if yours does. Bullet points are used to distinguish different points you wish to make, for example, in a summary:

My research project entailed:
- making use of irradiation fusion and microcell mediated chromosome transfer methods to place Hsa21 into an undifferentiated F9 cell
- treating two of the resulting cell hybrids, IY1 and IY5, with retinoic acid to induce neuronal differentiation
- creating a cDNA library from the treated IY1 cell line
- screening the library with a *Drosophila Mittnacht* cDNA probe
- isolating a human *Mittnacht* cDNA (*SM1*)

There are no rules as to whether bullet pointed items should finish with full stops, but they often look messy if they do (we do not use them in this guide).

Footnotes

Footnotes and endnotes are used to give additional information not strictly necessary to the argument. Footnotes are given at the end of the page *, endnotes are given at the end of the chapter, or in a list at the end of the book. Keep footnotes and endnotes to a minimum as they are distracting to read. If the information is important, include it in the main text; if not, think about whether you need to include it in your dissertation or thesis.

If you are using numbers to mark citations, it is best to use symbols * # § ¶ ‡ rather than numbers to mark your footnotes and endnotes, otherwise they might become confused with your reference citations. Put the symbol in superscript (above the normal line) so that it is not read as part of your main text.

---

*like this

Check Your Layout on the Final Version of Your Manuscript

Print one final version of your thesis or dissertation, and skim through this checking that all your formatting is consistent and you have no errors, such as leaving a heading at the bottom of a page, with no text underneath it.

Common Mistakes

- Inconsistent layout
- Not clearly separating paragraphs and other sections
- Excessive numbering of sub-sub-sub-sub-sub headings
- Leaving a heading as the last line of a page

Key Points

- Use a clear, easy to read font, such as 12 point Times or 10 or 12 point Helvetica
- Use one-and-a-half, or double line spacing (and check your local rules for thesis layout)
- Leave enough space on the left margin for binding
- Check the final layout on a printed version of your manuscript

---

*like this

Chapter 14

THE USE OF ENGLISH

To convey your meaning to the reader you need a good command of English. Bad English is not only irritating to read, but can also affect the meaning of what you write. In this chapter we start by looking at a few basic terms and constructions. We then look at scientific style, and go on to consider grammar, vocabulary, and punctuation. If you are seriously interested in, or worried about, your English, buy a good grammar book, usage book and dictionary.

A Few Basic Terms and Constructions

Subject

The subject of a sentence is the person or thing about which the verb expresses something, for example, in the sentence 'I am alive.' *I* is the subject (*am* is the verb).

Object

The object of a sentence is the person or thing that is acted upon by the verb in the sentence, for example, in the sentence 'David dropped the hydrochloric acid.' *the hydrochloric acid* is the object (*David* is the subject).

Nouns

These are names of things.

Common nouns, for example, *car, house, man, country, chemical, isotope,* are names for things in general.

Proper nouns, for example, *Boris Karloff, Paris, Edinburgh University, Renault 11,* are the names of specific people, places, organisations, or things.

Countable nouns, for example, *men, radioisotopes, eggs,* are names for things that can be counted. They have plural forms and we can write, for example:

Five men put two radioisotopes in eight eggs.

Uncountable nouns are names for things that cannot be counted, for example, *milk, oxygen, genetic engineering,* so we write,

Five men put two radioisotopes in some milk.

Collective nouns (also called compound nouns) are nouns that name a group of people, animals, or things, for example, *a flock of sheep, a committee of academics, a group of chemicals.*

Adjectives

Adjectives describe nouns, for example, *blue* and *fascinating* are both adjectives: 'the *blue* cheese', and 'the *fascinating* theory of relational databases'.

Verbs

Verbs are used to express a state of being or an action, for example, 'I *am*.', and 'The kangaroo *hops*.'

Transitive verbs require an object and can be either:

> active: I *ate* the cheese.
> or passive: The cheese *was eaten.*

Intransitive verbs do not require an object: Dr Karloff *walked.*

Auxiliary verbs are used to form a tense or an expression, for example, 'I *am* happy.' 'She *feels* sad.' 'I *have been* to Paris.'

Infinitives

An infinitive is a verb with *to* in the form, *to work, to have worked, to be working,* etc.

Participles

A present participle is a verb in the *-ing* form, for example, *thinking.*
A past participle is a verb in the form, *removed,* or *ran.*
Both types of participles can be used as adjectives:

> The *smiling* faces of the children.
> Recently *appointed* Dr Adams started his work on Landfish with blue plummage.

Gerunds

These are verbs in the *-ing* form, used as nouns, for example, *smoking*:

> *Smoking* is bad for you.

Active voice

This is when you use verbs in the following form: 'I *shot* the sheriff.', 'Dr Karloff *is parking* his car'.

Passive voice

This is when you use verbs in this form, 'The sheriff *was* shot.', 'Dr. Karloff's car *is being* parked.'

Person

By person we mean from what point of view you are writing:

First person:	*I, we, me, us*
Second person:	*you*
Third person:	*he, she, it, they, him, her, it, them*

Adverbs

Adverbs are used to describe a verb or an adjective, for example,

The charge-transfer complex was broken down *completely*.
The samples were *unusually* smooth.

Prepositions

Prepositions are words like, *in*, *of*, *next to*, etc. which show the relationship between things.

Pronouns

Pronouns are words like, *it, the, him, her*, etc. which you can use in place of nouns.

Conjunctions

Conjunctions (*and, but, therefore, however, or,* etc.) are used to join clauses together.

Tenses

Future simple

> The experiment will work. The experiment shall work.

We use this to express things we think or believe will happen.

Future continuous

> The experiment will be working...

This is usually used with a time expression for continuous actions before another point in time.

> The experiment will be working by the time Dr Karloff arrives.

Future perfect

> The experiment will have worked...

We use this to talk about things we think or believe will happen by a certain stated time in the future.

> The experiment will have worked, by the time I get back.

Present simple

> The experiment works.

We use this to talk about habitual actions, universal truths and opinions, and definite arrangements in the future.

Present continuous

> The experiment is working.

This is used to express things that are happening now, but not necessarily at the precise moment of speaking. We also use it to talk about definite arrangements in the near future:

> I am going to the cinema tonight.

Present perfect

> The experiment has worked.

This is used to talk about things that occurred in the near past, and for things that have happened in the past where no time is mentioned, for example:

> I have been to France [at some point in my life].

Present perfect continuous

> The experiment has been working...

This is normally used with a time expression for an action that started in the past and is still happening.

> The experiment has been working since I came back.

Past simple

> The experiment worked.

We use this to express things that happened in the past, past habits, etc.

Past continuous

> The experiment was working.

We use this for past actions that continued over a period of time. It is often used with the past simple to show that something was happening before a certain point in the past,

> The experiment was working until Dr Karloff turned off the apparatus.

Past perfect

> The experiment had worked.

This is used as a past equivalent of the present perfect. It is often used when a time is mentioned.

> The experiment had worked by 17.30.

Past perfect continuous

> The experiment had been working...

This is the past equivalent of the present perfect continuous and is normally used with a time expression.

> The experiment had been working for six hours by the time Dr Karloff noticed he had not removed the sodium crystal.

The imperative

Do not stick your fingers in the machine.

This is used when giving directions and orders.

The first conditional

If the experiment works I will/shall be famous.

This is used for things we consider probable.

The second conditional

If the experiment worked I would be famous.

This is used when we do not expect the thing to actually happen, and when what we are proposing is contrary to the known facts:

If astrophysics was easy, everyone would be studying it.

The third conditional

If the experiment had worked I would/could have been famous.

This is used for things that did not happen because something else did not happen.

Scientific Writing Style

One of the most important skills to learn as a scientist is how to communicate effectively and efficiently. Scientific writing style aims to be clear, precise and accurate. When writing scientific English, keep your prose impersonal because the reader is not interested in you *per se*, but in your research. Avoid emotive adjectives because if the research is *fascinating, interesting, relevant*, or *important*, you do not need to tell the reader so—it should be self-evident from the text.

Write in a straightforward style, that presents only what you need to say. Keep the words to a minimum and remove any that are not necessary, but make sure the text is easy to read and flows. If you get stuck, try pretending that you are telling a friend what you have done; use a normal conversational style and then change this into formal scientific writing.

Bear in mind that your dissertation or thesis is a serious scientific document, but do not try to force your writing into a style that feels uncomfortable or unnatural. Do not be tempted into using long impressive sounding words in an effort to make the writing sound more important and scientific. Quantity is no substitute for quality. Abstruse English is boring, pretentious, and dull, and if the words are used catachrestically (improperly) it looks silly as well. Avoid slang and other colloquialisms and, apart from in your Acknowledgements, there is no room for metaphor, irony, or jokes.

Which voice and person to use

One of the first things you will have to decide is which voice (active or passive) and person (first, second, or third) to use. Many people think that they have to use the third person passive to sound scientific:

It was found...

rather than the first person active:

I/We found...

The passive is used when we do not know or do not need to know who or what carried out the action, which is the case with a lot of science; however, the passive can become very wordy and pompous if used to excess, and there is no reason to conceal the fact that it was actually you who did your research. Our advice is to use the active as much as possible. Compare the following:

> Third person passive: The digests were continued for a further
> 5 hours...
> First person active: I continued the digests for a further 5 hours...
> Third person active: Digestion continued for a further 5 hours...

The third person active is shorter, more direct and easier to read. Whilst the third person passive is quite acceptable when writing about what you did:

> The blue mice were placed in the maze.

Try to avoid it when you are talking about what other people or things did. If you compare these two sentences:

> The maze was successfully traversed by the blue mice. (passive)
> The blue mice successfully traversed the maze.   (active)

Again, the active sentence does the job just as well and more efficiently.

Which tense to use

You will use a range of tenses depending on what you are writing about. It is best to use past tenses, on the whole, when talking about what you did, but beware of confusing the use of the present simple and the past simple, because this will affect the meaning of your sentence. Amongst other things the present simple is used to express habitual actions, 'Tim cleans his teeth every morning.', universal truths, 'Water boils at 100°C.', and universal opinions, '£100 is better than a slap in the face with a wet fish.' The past simple is used to say what happened in the past: use it for things that you did, 'I put the chemicals in the refrigerator.', observations you made, 'A shiny deposit of an unknown material appeared on my bread.', and specific rather than general conclusions you came to, 'I concluded that I should not eat the bread.'

Sometimes it does not matter which tense you use, for example, in your Materials and Methods you could write either '5 ml water are added' or '5 ml water were added'. In the first sentence you would mean, 'in the procedure 5 ml of water are always added'. In the second sentence you would mean, 'I added 5 ml of water when I did it'. If you are writing about the result of your experiment, be more careful, for example, '5 g gold are precipitated' means that it always is, '5 g gold were precipitated' means that it was in your experiment.

Comprehensibility

The following sentence, from a BSc thesis, is very hard to understand:

> The pouring of the gel was immediate as the tendency to set was great.

This probably means:

> The gel was poured immediately because it sets quickly.

Definitions

If there is any ambiguity or controversy about any of the words, phrases, or ideas you are using, define your terms so the reader knows how you are using them.

Directly addressing the reader

Your thesis or dissertation is a formal document so avoid phrases like 'You can see the progression in Figure 3.6.' If you have to address the reader, you could use either *we* 'We can see the progression in Figure 3.6.', or if you feel the need to be more formal, *one* 'One can see the progression in Figure 3.6.' However, it is far better, in a formal document, to avoid addressing the reader directly and to use the third person passive, 'The progression can be seen in Figure 3.6.'

Distinguishing between fact and opinion

If you are sure of a point in your argument, do not deflate its power by introducing it with phrases like 'In my opinion...', or 'It is possible that...'

But if you are dealing with points that you are less sure of, or which are just suppositions, let the reader know this—here phrases like 'It is possible that...' are appropriate.

Headings and titles

Choose short self-explanatory headings and titles to separate the different sections of your thesis or dissertation.

Latin names for organisms

Many scientific words are derived from Latin or Greek. They usually do not have an *s* on the end if they are plural, for example, data, bacteria. Non-English words are often italicised (see Italics section, below). Remember that Latin names for organisms are normally italicised. The genus name is capitalised and the species name is not, for example, *Caenorhabditis elegans* or *Mus musculus*.

Laws

Write the names of laws in lower case, for example, *the first law of para-psychology*, unless the laws contain a person's name, for example *Karloff's law*.

SI units

SI Units (Système International d'Unités) are the internationally standardised units of measurement. Use them in the correct abbreviated form and only capitalise them if they are normally capitalised. Whether you put a space between the number and the unit is up to you (*5kg*, or *5 kg*) but be consistent (see *Appendix 7*).

Numbers

In discursive scientific text, such as your Introduction or Discussion, it is normal to write out numbers from one to ten, for example, *two mice, nine peaches*. Fractions of two words can also be written, for example, *five-tenths of a Mars bar*. Larger numbers and fractions, or those written in methods and protocols, and attached to units (such as 28 g or 54 A) are shown as numerals. If you have two numbers together in a sentence, it is normal to write one of them in words, for example, *seven 5mm O-rings, fifty-eight 70 years olds*.

It is normal to write numbers in titles, unless it would be very unwieldy:

The Four Fosters
The Seventy Samurai
The Twenty-Five Favourite Sayings of Dr Karloff
The 3,599 Dalmatians

As far as possible in the discursive sections of your text, such as the Introduction and Discussion, do not start a sentence with a numeral (some disciplines disagree about this, see a good recent thesis in your field):

2 theories explain the phenomenon of Moog Genesis.

is better written:

Two theories explain the phenomenon of Moog Genesis

If you are numbering points in your text, put them in brackets. Roman or arabic numerals are equally acceptable, providing you are consistent:

There are a number of things to take into consideration when deciding on the verdict (1) the sausages were very tasty, (2) the sausages were left out on the kitchen table, (3) the defendant is a dog and cannot be held to be responsible for his actions.

There are a number of things to take in to consideration when deciding on the verdict (I) the sausages were very tasty, (II) the sausages were left out on the kitchen table, (III) the defendant is a dog and cannot be held to be responsible for his actions.

If you are writing out a series of numbers with units, for example, a series of measurements, avoid something like this:

12m, 13m, 14m, 17m...

It is much simpler to write them thus:

12, 13, 14, 17m

If you wish to show a span, such as 12, 13, 14, 15, 16, 17m, use an n-dash (see the section on punctuation):

12–17m

or write

12 to 17m.

A colon is used between numbers in proportions, for example, *5:1*.

Times of the day, days of the week, months and years

Times of the day should be written using the 24-hour clock, 22:13, rather than 10:13 pm. Days of the week are capitalised, for example *Wednesday*, not *wednesday*, and can be abbreviated without a full stop, *Wed*, for example. Months are capitalised, *September* not *september*, and can be abbreviated without a full stop, *Sept*, for example. Seasons, for example, *summer*, are not capitalised. Write years as numerals, *1998*, not in full, *nineteen hundred and ninety-eight*.

Referring to people

When writing about people avoid making generalisations that might cause offence. If you write 'When a nuclear physicist considers this problem he will come to a different conclusion to that of a marine biologist.' It should not need saying, but it has to be said, you are making an unscientific assumption that the nuclear physicist is male. This is likely to cause offence, and, apart from anything else, if your examiner is a woman you do not want to irritate her. You could get round the problem by writing he/she but this looks terrible. As much as possible avoid constructions where you are forced to specify a gender unless you know what the gender actually is. A simple way round the problem is to use the term 'they' to mean 'he/she'; this is an old construction not much seen in current English, but we think it is useful, 'When nuclear physicists consider this problem they will come to different conclusions from a marine biologists.'

Deflating your argument

Keep an eye out for constructions that deflate your argument. For example, if you use *definitely* too much you will imply that other points in your argument are not definite. If you write something like 'Personally I think...' you imply that other people would disagree with you. Of course if you are stating a personal opinion you should say so.

It is best to start a sentence strongly and then modify it as necessary, rather than starting with a long qualification:

> The result looks promising, although full analysis of the samples has not yet been done, and there is some contamination of samples, 3, 7, and 15.
> Although full analysis of the samples has not yet been done, and there is some contamination of samples, 3, 7, and 15, the result looks promising.

The second sentence is much less positive than the first and emphasises the qualifications, rather than that the results look promising.

Using nouns as adjectives

Nouns used as adjectives often have a clumsy feel: *apparatus construction*, although using less words than *construction of apparatus* feels ungainly and can be hard to read. This is not to say that nouns should never be used as adjectives, phrases such as, *hydrogen bond,* and *SI units*, are fine, but try to avoid writing things like *maximum permissible dose administration,* or *vegetative propagation measurement,* when you could write *administration of the maximum permissible dose,* and *measurement of vegetative propagation.*

Colloquialisms and contractions

Colloquialisms are words or patterns of speech that are used in everyday spoken English. Colloquialisms can easily lead to ambiguity, particularly if your text is being read by a scientist who works in a slightly different field or for whom English is a second language. For example, we recently saw the following in a dissertation:

> … the hot DNA probe was hybridised to the filters.

By 'hot' the writer meant radioactive, but there is no reason why the reader should know this. Be especially wary of using colloquial versions of standard abbreviations, such as referring to methanol as MeOH, as biologists do in the laboratory, rather than the standard abbreviation, CH_3OH.

Contractions such as, *it's, there's, I've, lab* (*it is, there are, I have, laboratory*) are colloquialisms which can very easily slip into your text. They should not be there, so run checks for them using the 'find and replace' command on your word processor.

Jargon

Depending on your point of view, jargon is meaningless gibberish, or the specialised language of a particular field. Use specialised terms that will be understood by your readers; but make sure you use them correctly, and that they are widely known and accepted.

Misuse of words

Be careful of everyday words that have a particular scientific usage, for example, *force, energy, inertia.* Use these words only in their precise scientific meaning. *Parameter* and *variable*, for example, are sometimes confused. *Parameter* means a constant that does not vary, as opposed to *variables*, which do vary. Remember that a datum is something given or assumed as a fact rather than preliminary results or findings—data are things given or assumed as facts. Also remember that *constantly* does not mean *often*, and that *efficient* does not mean *effective*, neither do *varying, variable, varied* and *various* mean the same thing.

Words with similar meanings or spellings are often confused, for example, *to accept* is to receive willingly, *to except* is to exclude or make an exception; *proceed* means to go ahead with, *precede* means to go before. Both are spelt correctly and neither will be picked up by a spell-checker. If you are using a word with alternative spellings, for example *extractable, extractible* make sure you are consistent. We have listed other problem words in *Appendix 1: Easily confused words*.

Typing mistakes can also lead to the wrong word being used. A spell-checker will pick up combinations of letters that do not form any word that it knows of, but your spell-cheekier will knot pique up wards that our correctly spelt butt wrong. It is easy to slightly mis-type a word and produce something you did not intend to, particularly words that are anagrams of each other, for example, *sorted* and *stored*. Make sure your text is properly proofread before handing it in.

Vagueness

Your examiners need to know exactly what you did and what you think. Chatting in the lab or over breakfast you might say things like 'The reaction was very quick.' If someone needs to know exactly how quick it was they can ask you. In your dissertation or thesis 'The reaction was very quick.' is irritatingly vague. *Quick* is a subjective idea, as is *very*. Is *quick* a nanosecond, or a few hours? Equally irritating are phrases like:

… less DNA was used.

How much is less DNA?

Having dried sufficiently…

What does this mean?

> An appropriate volume.

What is an appropriate volume? Also be careful about making imprecise comparisons, for example:

> The mauve mouse is heavier than the blue mouse.

Subjective adjectives and adjectival phrases like these should be banished from your text. It is a good idea to run checks for them while you are word processing. We have included a list of words and phrases that can cause vagueness in *Appendix 3: Wordy words and phrases*.

Wordiness

There are some words and phrases that always tend towards wordiness. Alarm bells should start ringing if you catch yourself writing things like:

> The use of gamma thobalot improved the patients' symptoms.
> The utilisation of gamma thobalot improved the patients' symptoms.

In both cases it would be more efficient to write:

> Gamma thobalot improved the patients' symptoms.

It is best, on the whole, to use plain rather than fancy words, for example, *start*, or *begin*, rather than *commence*; *attempt* or *try* rather than *endeavour* or *essay*.

In *Appendix 3: Wordy words and phrases*, we have listed words that can cause wordiness, but here are a few of the more common ones with their less wordy alternatives:

> ... boiled on a heating block so as to denature the protein.
> ... boiled on a heating block to denature the protein.
> Washing of filters.

Filter washing.
... the columns were left sitting for 2 minutes.
... the columns were left for 2 minutes.
the process of DNA preparation involves...
DNA preparation involves...

That can also lead to wordiness. It is used far more often than is necessary, for example, in the sentence:

There are three things that this depends on...

that is not necessary. You could just write:

There are three things this depends on...

Keep an eye out for tautologies—using more than one word to express the same idea in the same sentence or phrase (for example, *attach... together, a pair of twins, seems that... could be possible, repeat... again, basic essentials*). The following sentence contains a tautology:

The reason my experiment failed is because I forgot to turn on the apparatus.

'The reason' and 'because' are both doing the same job; write either:

The reason my experiment failed is I forgot to turn on the apparatus.

or, more succinctly:

My experiment failed because I forgot to turn on the apparatus.

You need to be precise and accurate in your choice of words. If there is an appropriate scientific term, use it.

Use distinctions and definitions where they are needed to avoid confusion, but try not to make them unnecessarily. We have bracketed the unnecessary distinction and definition in the passage below:

The Yellow Wagtail is a British sub-species of the Continental Blue-headed Wagtail. It is a summer visitor [and migratory bird of passage], mainly found in pastures [rather than woodlands, marshes, or mountains].

Prepositions

Avoid unnecessary prepositions, such as 'The thesis will be printed out' rather than 'The thesis will be printed'.

Scientific writing often omits prepositions when describing methods or results, so we would write '10g $CuSO_4$' rather than '10 grams of copper sulphate'. We would also write '5ml water were added rather' than '5ml of water were added', or '10nm diameter', rather than '10nm in diameter'. Read published papers and learn to use the conventions of your field so that you can produce a professional looking thesis or dissertation.

Repeating words

Unless you are consciously doing so to stress or make a point, it is best not to use a word too many times in the same paragraph or sentence. It will not affect the meaning of your text, but is tedious to read. Buy yourself a thesaurus—you might have one as part of your word processing program. If you find yourself using the same word again and again, have a look in your thesaurus for an alternative. Before you use an alternative, check what it means. It might only have a similar meaning, which may distort what you are trying to say, or just sound odd:

> ... While you should use distinctions and definitions where they are needed to avoid confusion, you should snub unnecessary distinctions and definitions...

American English and British English

Be aware of the differences between American and British English, most of which are to do with spelling, for example, *grey*, *utilise*, and *honour* are standard spellings in British English, but *gray, utilize,* and *honor* are standard spellings in American English. You could use either spelling system providing you are consistent. If you are studying at a British university it is sensible to use British English. Make sure your spell-checker is set appropriately.

Literary quotations to preface chapters

Some authors like to put a literary quotation at the start of each chapter. There is no rule against doing this in your thesis, but we do not recommend it as many examiners frown on the practice. We have seen an extreme example in a PhD thesis that not only had literary quotations, but also little illustrations and cartoons in the margins of the first page of each chapter rather like mediaeval illuminations. You would certainly never be failed for this type of style, but not all examiners share your sense of humour. If you do use literary quotations, then make sure that you acknowledge the source of the quotation, and make sure that this acknowledgement is correct.

Planning Your Writing

Just as you should plan the grand sweep of your argument it is a good idea to plan how to present your information, from paragraph to paragraph, and from sentence to sentence. Make sure your ideas flow rationally and understandably from one sentence to another and from one paragraph to another.

Planning your paragraphs

The structure of your paragraphs should mirror that of your whole text. They should have a beginning, a middle, and an end. Restrict each paragraph to one main idea, which you should give to the reader in your 'topic sentence'. The topic sentence will usually come at the beginning of the paragraph and tells the reader what the paragraph will be about. Keep this topic sentence in mind as you write the rest of your paragraph to ensure that the paragraph is coherent within itself. Each sentence should flow logically on from the previous sentence to build your argument towards the end point of your paragraph. If you find you need to include ideas which are not covered by your opening topic, either widen your topic to encompass them, or put these ideas into another paragraph.

When planning how to write a chapter or section, list the topics and arrange them into a rational order. Once you have the topics in order, you have the backbone of your writing in place. You can now go through this outline noting down the points you wish to discuss under each topic. These will form the core of your sentences. Make sure that your sentences build rationally to a main point at the end of each paragraph.

Make sure your writing flows

You might find that the sentences in your paragraph, although all connected to the topic sentence, seem to jump about rather jerkily from idea to idea. You can use transitional words and phrases like *therefore, in particular, but, although, since, because,* etc. to lubricate the changes and show the reader where your argument is going.

If you find that, while they work well in themselves, your paragraphs do not seem to fit very well together, a useful trick is to simply repeat a word or idea from the conclusion of the previous paragraph as an introduction to the next one to show the transition, for example:

> When the pistons move up and down the connecting rods make the offset crank pins rotate, thus causing the crankshaft itself to rotate. We see therefore that crankshafts are an essential component in the piston engine.
>
> Another interesting component of the piston engine is the piston itself...

Dealing with awkward points and comparisons

Deal with counter-arguments and alternative interpretations immediately. You do not want to give them a chance to take hold and distract the reader's mind from your argument. With more complicated counter-arguments, you should break them down and deal with them point by point. If you just lay out a counter-argument in full followed by your refutation of it, the reader will find it hard to follow and possibly lose track of what you are saying. Similarly, if you are comparing one quite complicated idea or process with another, it is a lot easier to grasp the comparison if you run the comparisons side by side, point by point.

Plagiarism

Never copy anything from papers or other people's work without fully acknowledging and referencing it. Students are failed for plagiarism.

Grammar

Just about everybody makes slips in grammar when they are talking. This is normal and forgivable. However, when it comes to the written word, particularly in a formal document like a dissertation or thesis, we expect precision and correctness. Bad grammar is not only irritating to read:

> The physicists, what play the piano, is in the next room.

it can also affect the meaning of what you write:

> The dashing young cosmologist Dr Anderson ate an apple roller-blading in Hyde Park.

Here it is unclear who was roller-blading, the dashing young Dr Anderson, or the apple. Better construction would clarify the meaning of this sentence.

> The dashing young cosmologist Dr Anderson ate an apple while roller-blading in Hyde Park.

In writing, unlike speech, you have time to revise and consider what you are saying, make use of this time.

Sentences

A sentence should contain at least one subject and predicate. The subject is the person, people, thing, or things the sentence is about. The predicate is something that is said about the subject. In the following sentence:

> I want to go to sleep.

I is the subject and *want to go to sleep* is the predicate. In the next sentence:

> John, Glenn, Tony and Lizzy are researching the effect of carbon monoxide on children's health in Glasgow.

John, Glenn, Tony and Lizzy is a 'compound subject', *are researching the effect of carbon monoxide on children's health in Glasgow* is the predicate. Both of these sentences are simple sentences. A compound sentence is made of two or more simple sentences, for example:

> I want to go to sleep but I have to stay awake.

Complex sentences are made by taking one sentence as a main clause, and adding other sentences in the form of subordinate clauses, for example:

> I want to go to sleep, because I am tired, but I have to stay awake.

because I am tired is the subordinate clause.

Sentence length

There are no rules as to how long a sentence should be. Short sentences tend to emphasise a point, but if you always write in short basic sentences your style will come across as abrupt and jerky and you will often have to repeat words and information in order to make your sentences grammatically correct. On the other hand, the longer a sentence is, the harder it can be to follow the meaning unless it is well punctuated—and the easier it is to write a bad sentence. If your sentence is longer than three or four lines, it is probably too long.

Relative clauses

Relative clauses are made using the relative pronouns, *who, which, that,* etc. There are three kinds of relative clauses, defining (or restrictive) relative clauses, non-defining (or non-restrictive) relative clauses, and connective relative clauses.

Defining relative clauses

These are used to define the thing we are talking about. For example, in the sentence:

> I put the adults *which I had covered with yeast extract* in a test tube.

which I had covered with yeast extract is a defining relative clause. It tells the reader which adults we are talking about. With defining relative clauses we do not use commas and can use the following relative pronouns:

For people:
Subject
We use *who* (*that* can be used after *all, everyone, everybody, no one, nobody,* and *those*):
> The woman who arrived yesterday is an engineer.

Object
That is normally used, although *whom* is used in formal English. *Who* can also be used:
> The man that/whom/who I saw is a geologist.

Possessive
Whose:
> The woman whose car is being towed is an astrophysicist.

For things:
Subject
We use *which* (formal) and *that*:
> The book that fell on my foot is heavy.

Object
We can use either *which* or *that*:
   The book which I dropped on my foot is heavy.
Possessive
Whose, or *of which* (formal):
   The book whose dust jacket I tore is my own.

With defining relative clauses we do not have to use relative pronouns, and could just write:

   The man I saw is a geologist.

You can also replace the relative clause with an infinitive in certain situations:
   After *the first, the second… the last,* etc:

   The last experiment that was done
   The last experiment to be done

and when you are talking about purpose or permission:

   I have a lot of experiments that I have to do
   I have a lot of experiments to do.

   Dr Karloff told me that I had to be here by eight o'clock to take part in the Moog Genesis experiment.
   Dr Karloff told me to be here by eight o'clock to take part in the Moog Genesis experiment.

You can replace the relative clause with a present participle when you are talking about: continuous or habitual actions, desires, knowledge, or thought:

   The lens which moves on the axis…
   The lens moving on the axis…

   The notice which warns people against touching the apparatus had been removed by Dr Karloff.
   The notice warning people against touching the apparatus had been removed by Dr Karloff.

Non-defining relative clauses

In non-defining relative clauses, for example:

> I put the adults, *which I had covered with yeast extract,* in a test tube.

we are simply give the reader more information about the thing we are talking about. The clause is put within 'parenthetic commas'—if the clause is removed the sentence should still be understandable. With non-defining relative clauses we have to use commas and use the following relative pronouns:

For people:
Subject
We use *who*:
> The woman, who arrived yesterday, is an engineer.

Object
Whom is formal, *who* is less formal:
> The man, whom/who I saw, is a geologist.

Possessive
Whose:
> The woman, whose car is being towed, is an astrophysicist.

For things:
Subject
We use *which*:
> The book, which fell on my foot, is heavy.

Object
We use *which*:
> The book, which I dropped on my foot, is heavy.

Possessive
Whose, or *of which* (formal):
> The book, whose dust jacket I tore, is my own.

With non-defining relative clauses you must use both commas and relative pronouns. If we compare the following sentences:

The chloride samples, which I left in my office, were ruined.
The chloride samples which I left in my office were ruined.

In the first sentence (with a non-defining clause) all the chloride samples were left in the office and ruined. In the second sentence (a defining relative clause) only those chloride samples which were left in the office were ruined, the other chloride samples were not left in the office and were not ruined.

You can replace the relative clause with a present participle when you are talking about: continuous or habitual actions:

The lens, which moves on the axis, has developed a crack.
The lens, moving on the axis, has developed a crack.

and when you are writing about expressions of desire, knowledge or thought:

Dr Karloff, who wanted to be alone, left the room.
Dr Karloff, wanting to be alone, left the room.

Dr Karloff, who knew he was being followed, leapt into the alleyway.
Dr Karloff, knowing he was being followed, leapt into the alleyway.

Dr Karloff, who thinks he has been treated badly, has been waging a vendetta against the University establishment.
Dr Karloff, thinking he has been treated badly, has been waging a vendetta against the University establishment.

Connective relative clauses

These are very similar to non-defining relative clauses, and are used to carry on a story:

Dr Karloff hit the Dutchman. The Dutchman fell to the ground.

can be combined:

Dr Karloff hit the Dutchman, who fell to the ground.

who fell to the ground is a connective relative clause. As with non-defining relative clauses they are indicated by commas and you must use a relative pronoun. If you write:

Dr Karloff hit the Dutchman who fell to the ground.

you mean Dr Karloff hit the Dutchman who had already fallen to the ground *who fell to the ground* becomes a defining relative clause. We use the same relative pronouns as for non-defining relative clauses:

For people:
Subject
We use *who*
Object
Use *whom* (formal), or *who*.

For things:
Subject
We use *which*.
Object
We use *which*.

Common problems

Unnecessary changes of tense or voice in mid-sentence

At times you will have to change tenses in the same sentence to get the correct meaning across, for example:

By the time the apparatus was set up, the chemicals had been prepared and were ready for use.

Here we shift from past to past perfect, which is quite normal. But avoid shifting tenses if it is illogical, or unnecessary:

The purple mouse goes through the maze and learnt that there was a piece of cheese at the end.

Fig. 14.1   The purple mouse goes through the maze and learns that the cheese has gone.

We have shifted from the present to the past tense which makes a nonsense of the sentence.

> As I continued with the experiment, the formation of gold deposits was observed.

Here we have shifted from the active to the passive for no good reason.

Verbs and subjects that do not agree

You have to make sure that the verb you are using is in the right form, or 'agrees with', the subject:

> Careful consideration of all the examples are needed.

'are' does not agree with 'consideration'. This should be written:

> Careful consideration of all the examples is needed.

In the following sentence *were* does not agree with *the purple mouse*:

The purple mouse, along with the orange mouse, the red mouse, and the blue mouse, were accidentally released into the wild.

This should be written:

The purple mouse, along with the orange mouse, the red mouse, and the blue mouse, was accidentally released into the wild.

In British English, collective nouns like *committee, group, class,* can be treated as either singular or plural:

The committee was dismissed.
The committee were dismissed.

American English normally treats them as singular.

Each, either, neither, anyone, anything, everyone, everything, no one, nothing, everybody, nobody are all singular so:

Each of the experiments succeeded in their objective.

is wrong—*each* is singular, *their* is plural. The sentence should read:

Each of the experiments succeeded in its objective.

If you are comparing a singular with a plural, the verb will match the one it is closest to:

Either the apparatus or the students are at fault.
Either the students or the apparatus is at fault.

People often have problems with the word *data*, which is not a collective noun, although it is often wrongly treated as one in everyday speech. *Datum* is singular. *Data* is plural. Write, for example, *the data are..., the data show..., the data were analysed...,* rather *than the data is..., the data shows..., the data was analysed...* Similarly, *media* is plural and *medium* is singular, *phenomena* is plural and *phenomenon* is singular.

Using the wrong pronoun

Cindy and myself carried out the experiments.

You would not write *Myself carried out the experiments*. The sentence should read:

Cindy and I carried out the experiments.

Here is another example:

The police are accusing Dr Karloff and I of masterminding the Tokyo Affair.

accusing I? Write:

The police are accusing Dr Karloff and me of masterminding the Tokyo Affair.

Not completing comparisons

The blue mouse is larger.

than what? You have to complete the comparison:

The blue mouse is larger than an elephant.

Also ensure your comparisons make sense, in the sentence:

The gauze absorbs more water than paper.

It seems as if you are saying that the gauze absorbs some paper but a lot more water. It would be clearer to write:

The gauze absorbs more water than the paper does.

Dangling participles

These are also known as hanging, unattached, and misrelated participles. If they dangle they do not agree with anything, or agree with the wrong subject. This does not normally cause misunderstanding but it can at times, and it is always jarring. For example, if someone was writing about an experiment they did using orange mice:

> While conducting the experiment the orange mouse became confused.

Fig. 14.2   While conducting the experiment the orange mouse became confused.

This means the orange mouse was conducting the experiment. The participle 'conducting' has to agree with a subject. Here the only subject in the sentence is *the orange mouse*. It would be better to write something like:

> While I was conducting the experiment, the orange mouse became confused.

There are a number of participles that are allowed to dangle free, as it were, for example, *assuming, considering, excepting, providing, supposing*. Strictly speaking we should write:

> Considering the recent crisis in the Computing Department, we think Dr Lugosi should resign.

Due to many centuries of usage it is not considered necessary and we can just write:

> Considering the recent crisis in the Computing Department, Dr Lugosi should resign.

The split infinitive

The split infinitive is one of the great shibboleths of English. An infinitive is a verb in the form, *to-*, for example, *to go*. To split the infinitive is to put an adverb (a word that describes a verb, such as *boldly*) between the *to* and the verb, for example, *to boldly go* rather than the more orthodox *to go boldly*. Some people say that to split an infinitive is a sin, others are more relaxed about it (including the Oxford English Dictionary). Whatever your views about split infinitives, be careful about using them in your dissertation or thesis. They are more acceptable in spoken than written English. Your dissertation or thesis is a formal document so use them sparingly, if at all. It is possible that your examiners will consider them bad English.

Where to put prepositions

In formal English we put the preposition before the noun or pronoun,

> *At* what are you looking?
> The girl *with* whom I am in love.

In informal English the preposition moves to the end of the sentence,

> What are you looking *at*?
> The girl I am in love *with*.

As your thesis or dissertation is a formal document it is best to stick to formal English as much as possible. With phrasal verbs, such as 'looking after', you cannot put the preposition before the noun or pronoun, *'After the experiment which I was looking'*, because the sentence then makes no sense, so the preposition has to go afterwards, 'The experiment I was looking *after'*.

Punctuation

Punctuation can be compared to the stresses, gestures, intonations and pauses that we use in spoken English to make ourselves understood. One usually does not notice good punctuation, but bad punctuation jars and interrupts the flow of your text. It can also affect the meaning of a sentence. Punctuation should be used to make it perfectly clear to the reader the meaning and sense of the writing and to guide them through it with the minimum possibility of ambiguity. It would be comforting to think that punctuation is governed by a few simple and logical rules. It is not. Punctuation is governed by rules, conventions, and taste in roughly equal measures. There is sometimes little logic to it. Generally, the shorter and simpler you make your sentences, the less problems you will have with punctuation.

Paragraphs

A lot of people are not quite sure exactly what a paragraph is—or should be. The word paragraph comes from the Greek *para* (a sheet of handwriting) and *graphos* (mark). Originally it was merely a mark in a text to indicate to the reader that they had reached a pause or change of tack in the argument, much as a full stop is there to indicate the end of a sentence. If you play around with your word processor you can see these marks on your screen ¶.

There are no rules as to how long a paragraph should be. It depends on the rhythm of your argument and on what looks good on the page. Newspaper paragraphs tend to be short and punchy, which gives a sense of energy and looks readable in the columns. In your thesis you do not want quite such a frenetic rhythm. Your text would read like a runaway lawnmower. Lay your points before your audience with a steady yet varied pace. A short paragraph tends to emphasise the point you are putting across. The pause between paragraphs gives the reader the chance to reflect on and be amazed by your insight. As a general guide, if you are writing in double-spaced type and your paragraph is longer than a side of paper, it is too long. A full page of unbroken text is a daunting sight. The reader is likely to lose the flow of your argument as they wonder where the next break comes.

Punctuation marks

Sometimes punctuation marks affect the meaning of a sentence, sometimes they help the flow of your text, and sometimes they just litter the page like mouse droppings on a shiny new floor. Use the minimum punctuation marks necessary to make your meaning absolutely clear.

Perhaps surprisingly, there are no rules as to how many spaces to leave after punctuation marks. It is conventional to leave a single space after most of them, and two spaces after a full stop to indicate a new sentence.

The comma

The comma is possibly the most misused of punctuation marks. Sometimes the use of a comma will change the meaning of your sentence, sometimes it will affect the precise shade of meaning, at other times it is a matter of taste. Used well, commas make it absolutely clear what you mean and give your writing a good rhythm. But, if they are overused or used incorrectly they can make your text look fussy and can cause confusion. We use commas in the following ways:

Commas and conjunctions

Conjunctions are words like, *and, but, therefore, so,* used to join clauses together. If you use a comma there is a break between the ideas and the sense of the sentence changes:

Dr Karloff has gone to Amsterdam or Upminster.
Dr Karloff has gone to Amsterdam, or Upminster.

In the first sentence it is equally likely that Dr Karloff went to either place. In the second sentence *or Upminster* is an afterthought we do not think as likely as the first possibility. It is more common, but not necessary, to use a comma with conjunctions that indicate a break between ideas:

> The monster was dead and it did not move when Dr Karloff touched it.
> The monster was dead, but it moved when Dr Karloff touched it.

You cannot put a comma between two sentences without using a conjunction, for example:

> Dr Karloff has disappeared, he was last seen catching a taxi to Heathrow airport.

The sentence should read:

> Dr Karloff has disappeared and he was last seen catching a taxi to Heathrow airport.

or,

> Dr Karloff has disappeared, and he was last seen catching a taxi to Heathrow airport.

Commas are also used to avoid ambiguities, for example, *however* has more than one meaning. *However* can mean either *nevertheless* or *in whatever way*. You have to be very careful about where you put your commas as it can change the meaning of your sentence:

> Dr Karloff went to the hotel in Berkeley Square, however Sergeant Piddock was waiting for him.

Here *however* means *in what ever way* (sitting in the room, hiding in a tree, disguised as a penguin, etc. which does not make sense). In the following sentence:

> Dr Karloff went to the hotel in Berkeley Square, however, Sergeant Piddock was waiting for him.

However means *nevertheless*. You need the comma after *however* to make your meaning clear. *Still* has similar problems. *Still* can mean *not moving, nevertheless*, or it is used to indicate that something is continuing:

> Still I found the exercise useful.

without a comma this either means that you were still whilst finding it useful, or that you continued to find it useful.

> Still, I found the exercise useful.

means 'Nevertheless, I found the exercise useful.' There are other times when a comma is needed to get the right meaning across:

> I turned and examined the samples.
> I turned, and examined the samples.

In the first sentence the samples were turned and examined. In the second sentence you turned and the samples were examined. The commas shows us that *turned* agrees with (goes with) *I*. Keep an eye out for ambiguities like this.

Commas with a list or series of phrases

You can use commas to separate the items in a list or a series of phrases:

> I went to the market and I bought a fat hen, a haddock, an overcoat, a lithograph of the Eiffel Tower.

You could use an *and* before the final item in the list to soften the feel a bit:

> I went to the market and I bought a fat hen, a haddock, an overcoat, and a lithograph of the Eiffel Tower.

People disagree about whether or not to use a comma before the final *and*, but it is safer to use one to avoid ambiguities like:

I went shopping at Harrods, W. H. Smith, Woolworths, Boots and Marks and Spencers.

You can use commas to separate a series of phrases in the same way:

In this dissertation I intend to investigate the relationship between air pollution and landfish population, the relationship between water borne pollution and landfish size, and the relationship between stress and abnormal reproductive behaviour in landfish.

Commas with introductory or inserted words and phrases

Often a comma is not strictly needed, but makes the meaning easier to follow by putting the word or phrase in parenthesis:

Finally, we realised we had failed to turn on the centrifuge.

If there is a possibility of an ambiguity, however slight, use a comma to make your meaning clear:

Whatever the position of the apparatus I found the rain in Spain fell mainly on the plain.

here it is a possible that you are writing about a piece of apparatus that you found.

Whatever the position of the apparatus, I found the rain in Spain fell mainly on the plain.

makes the situation clear.

Commas and relative clauses

We use commas with non-defining relative clauses (see the grammar section):

The blue mouse, *which is very intelligent,* has escaped and is believed to be responsible for the recent power cut.

and connective relative clauses:

The blue mouse, which is very intelligent, has escaped and is believed to be responsible for the recent power cut, *which has thrown the university into chaos.*

but not defining relative clauses:

The blue mouse *which is very intelligent* has escaped and is believed to be responsible for the recent power cut.

Commas with *if* clauses and *when* clauses

In a sentence like:

You will be ill if Dr Karloff puts the chemical in your tea.

or,

She was ill when Dr Karloff put the chemical in her tea.

we do not need a comma, although it can be useful to use commas to guide the reader through a particularly long and complicated sentence. It is, however, common (not necessary) to use a comma when the if or when clause comes first:

If Dr Karloff puts the chemical in your tea, you will be ill.
When Dr Karloff put the chemical in her tea, she was ill.

Inserting words, phrases or clauses into sentences

If you are inserting a clause, phrase or word into the middle of a sentence with commas you must have both an opening and a closing comma just as you would use opening and closing brackets. A very good way of checking to see if you have used the commas correctly is to imagine replacing them with brackets:

> Dr Karloff braced himself and, cursing Carter, jumped from the open door of the cargo plane.
> Dr Karloff braced himself and (cursing Carter) jumped from the open door of the cargo plane.

> The Dean, Dr Moon, was arrested for drink-driving outside Saltdean Lido.
> The Dean (Dr Moon) was arrested for drink-driving outside Saltdean Lido.

E.g., i.e., viz., etc.

We use a comma before and a comma or a colon after words or phrases like: e.g., for example, i.e., that is, and viz. to put them in parenthesis:

> When served, fried landfish often tastes better with the addition of seasoning, for example, salt or peper.

Commas and adjectives

Do not put a comma between the last adjective and the noun:

> Dr Karloff held up the bright, sparkling, iridescent gem-stone and sighed with satisfaction.

If the adjectives are closely associated forming a kind of compound adjective, the commas are left out, for example, an *air cooled rotary engine*.

Commas and dates

In British English commas are not put around the year:

> *July 18 1960*

in American English they are:

> *July 18, 1960*

The colon and the semi-colon

The colon

We use the colon between sentences when there is a movement forward in the ideas:

> The chemist and physicist Michael Faraday started work as a laboratory assistant to Sir Humphrey Davey in 1813, in 1821 he began experimenting on electromagnetism, and in 1833 succeeded Davey as professor of Chemistry at the Royal Institution: he went on to become one of the foremost scientists of his day.

We use the colon to introduce examples and quotations:

> … many attempts have been made to overcome this problem, for example, Karloff maintains:
>> The normal phases of catabolism, in which complex substances are decomposed into simple ones, and anabolism, in which complex substances are built up from simple ones, have been bypassed by the technique of Moog Genesis…

We can also use a colon instead of a comma after phrases like: e.g., for example, i.e., and viz.:

> There are two things the Big Bopper likes, viz.: Chantilly lace, and a pretty face.

We use the colon before lists of things, for example:

> I went to the market and I bought: a fat hen, a haddock, an overcoat, and a lithograph of the Eiffel Tower.

Using a colon rather than a comma makes it more list-like. When you are introducing your list with phrases such as *the following, are these, namely,* always use a colon:

> I went to market and I bought the following: a fat hen, a haddock, an overcoat, and a lithograph of the Eiffel Tower.

We also use the colon before a quotation:

> Dr Cohen maintains that:
> A free radical acts like a drunk in a midnight choir and has often been compared to a bird on a wire...

and before a summary of a situation:

> The conditions under which I was working changed considerably: my collaborator Clarissa had left to take up a career on the stage, my supervisor Dr Karloff had disappeared (he is rumoured to be in Amsterdam), and I was having severe doubts as to the point of my research.

The colon is also used between numbers in proportion, for example, *5:1,* and can be used when writing times of the day, for example *21:30*.

The semi-colon

You can use semi-colons to separate clauses which are of more or less equal importance:

> Dr Karloff has disappeared; he was last seen catching a taxi to Heathrow airport.

Semi-colons are not normally used with conjunctions, but can be if you want a greater break than a comma would provide.

> Dr Karloff has disappeared; and he was last seen catching a taxi to Heathrow airport.

When using a semi-colon in this way you have to make sure that each clause could stand as an independent sentence. You could not write:

> Dr Karloff has disappeared; and was last seen catching a taxi to Heathrow airport.

was last seen catching a taxi to Heathrow airport does not work on its own as a possible sentence as it does not have a subject—use a comma. Generally, we use semi-colons with conjunctions that co-ordinate (give equal importance to) *and, or, but, however*, etc. but not with conjunctions that sub-ordinate (give a lesser importance to) *as, since, because*, etc.

We can also use semi-colons instead of commas to separate a series of phrases. They are especially useful when the phrases themselves contain commas:

> The University of West Cheam has recently been involved in a series of unfortunate scandals; the arrest of Dr Karloff on drugs and pornography charges; the resignation of Vice-Dean Martin over the Tokyo Affair; the fabrication of student numbers, which only came to light last week, in the Para-Psychology Department; and the failure to keep up mortgage re-payments for the Salisbury Avenue site.

The full stop

The full stop (period if you use American English) is used at the end of a sentence. We can also use the full stop when we are laying out lists in the following style:

> Common mistakes:
> Forgetting to include citations in the references.
> Using incorrect journal abbreviations.
> Inconsistency of style.
> Mis-spelling authors names.
> Getting the page numbers wrong.

You do not have to use full stops and could write:

Common mistakes:
Forgetting to include citations in the references
Using incorrect journal abbreviations
Inconsistency of style
Mis-spelling authors names
Getting the page numbers wrong

But, in British English, we should not use them like this:

> Dr Karloff's text included many common mistakes: Forgetting to include citations in the references. Using incorrect journal abbreviations. Inconsistency of style. Mis-spelling authors names. Getting the page numbers wrong.

Here, semi-colons or commas should be used.

The full stop in abbreviations and contractions

In contractions, when we are using simply the first and last letter of the word, full-stops are not commonly used: *Mr*, *Mrs* and *Dr* for *Mister*, *Mistress* and *Doctor*. Note that *Ms* is a recent invention and is not an abbreviation for any word and so should not have a full stop after it.

In abbreviations, where we are using just the beginning of the word, it is more common to use the full stop: *Prof.*, *etc.*, and *et al.* for *Professor*, *et cetera*, and *et alia*.

Increasingly full stops are being abandoned altogether in abbreviations and contractions, particularly the more common ones: *TV, UK, MI5, BBC, 1000 BC, 11 am, 45 rpm, 3500 kwh, MSc, PhD, DPhil*. However, use a full stop if there is a possibility of ambiguity. Convention still dictates the use of full stops with some common abbreviations such as *e.g.* (*exempli gratia*, for example), *i.e.* (*id est,* that is), *etc.* (*et cetera*, and so forth), *et al.* (et alii, and others), or *viz.* (*videlicet*, namely). Do not use full stops with the abbreviations of SI units.

Whether you choose to use full stops or not in abbreviations, make sure you are consistent. If you are not sure whether or not to use a full stop, err on the side of caution and use one. We came across one student who had not put a full stop after *et al.*, and was made, by a particularly pedantic examiner, to correct his thesis by putting a full stop after every *al* in the text.

If you are using a question mark after an abbreviation with a full stop, put the question mark after the full stop, *Have you got a Ph.D.?* This also holds for exclamation marks, *He has not put a full stop at the end of et al.!* Do not use exclamation marks in your text!

If your sentence ends in an abbreviation with a full stop do not put another full stop in, simply end the sentence with the abbreviating full stop.

If you are using an abbreviation in your text, it is common practice to write the word in full the first time it is used with the abbreviation in brackets after it, and then to simply use the abbreviation—do not do this with standard abbreviations such as SI units, chemical symbols, etc.

Fig. 14.3   A quiet night at the rat and maze.

Full stops in titles and headings

Do not use a full stop at the end of a heading or title.

Brackets

In a written text (we are not going to consider their use in equations and formulae) rounded brackets (or parentheses) are used (like a pair of commas) to separate a number, word, phrase, or clause from the main sentence. Use them if you are making numbered points in your text:

> We will look at (1) the behaviour of the purple mollusc before treatment, (2) the behaviour of the purple mollusc during treatment, (3) the behaviour of the purple mollusc after treatment.

You can also use them when you want the effect to be more of an aside than a pair of commas produce:

> The University of West Cheam (formerly West Cheam Technical College) has been closed down temporarily.

Do not let a pair of brackets disturb the normal punctuation of the sentence. In the following sentence we need the final full stop:

> The data from the final experiment were inconsistent (see Fig. 6).

We normally only use brackets around short phrases, clauses or sentences we are slipping into a main sentence. Although it is possible to use brackets around whole sentences or groups of sentences it looks very unwieldy. (If you do bracket an entire sentence, the full stop goes inside the brackets, like this.) Square brackets [] are used mainly with quotations either to fill in some missing information:

> The [principal] aim of the Genetic Engineering Department is to produce blue dandelions.

or with *sic* (so) to indicate a mistake in the quotation:

We normally put soddum [sic] chloride and vinegar, ascetic [sic] acid, on our fish and chips.

Curly (or hooked) brackets {} can come in handy if you are using brackets within brackets as it were, {()}, but if your text is getting that complicated you are using too many brackets and should think about re-writing your sentence. Angle brackets <> are not normally used in written text.

Dashes and hyphens

There are two sorts of dashes: m-dashes and n-dashes. The m-dash (the width of an m) is slightly longer than the n-dash (the width of an n). Hyphens are slightly shorter then n-dashes—all very confusing. You will be able to produce the different marks by playing around with your keyboard. On our keyboard a hyphen is produced by pressing the hyphen key -, an n-dash is produced by pressing the option and the hyphen key together –, and an m-dash is produced by pressing the shift, option and hyphen key together —. Do not put spaces between a hyphen and the words it joins. Opinion is divided as to whether or not to put spaces between an m- or n-dash and the words it separates—use whichever convention you feel most comfortable with.

The m-dash

The m-dash can be used to insert extra information into a sentence in the same way as commas and brackets:

> The University of West Cheam—formerly West Cheam Technical College—has been closed temporarily.
> The data from the final experiment were inconsistent—see Fig. 6.

You can use an m-dash like a comma before an afterthought—the effect is more abrupt and immediate:

> Multiplying by 5 brings us to the answer—at a price.

You can also use it like a colon:

> The conditions under which I was working changed considerably—my collaborator Clarissa had left to take up a career on the stage, my supervisor Dr Karloff had disappeared (he is rumoured to be in Amsterdam), and I was having severe doubts as to the point of my research.

In this example we have used brackets rather than dashes to separate *he is rumoured to be in Amsterdam* from the sentence, as further use of dashes would be confusing.

Apart from after a question mark or exclamation mark never use dashes or hyphens next to another punctuation mark. Dashes are not an all-purpose substitute for other punctuation marks. They can liven up your text but do not be tempted to overuse them or your text will look like Morse code.

The n-dash

You use the n-dash to indicate a span of space or time, and page numbers:

> The London–Brighton road.
> 1972–1985
> pp 3–17

The hyphen

Like the comma, the hyphen is a small and often misused punctuation mark. Hyphens are used to join two words together to make a new word, for example, common-sense. Whether or not words are hyphenated depends on the stage they have reached in their evolution. With fairly recent pairings we tend to write them as two separate words. When they have become reasonably accepted they progress to being hyphenated. Once they have become generally accepted we write them as one word, for example, handkerchief. Although they have been around for a long time, a few words like *common sense / commonsense / common-sense* and *good bye / goodbye / good-bye* still have not settled down into one standard form or another. If you are not sure about a word, be guided by your spell-checker (and dictionary), which will at least be consistent. There are certain times we always use hyphens:

With pure prefixes:

Ex-President, Vice-Dean, Sub-Postmaster

With proper names:

anti-French, un-American, pre-Christian

With suffixes to single capital initials, symbols, Greek letters, etc.:

X-ray, ß-ray

To avoid double 'i's, 'o's and triple consonants:

anti-intellectual, bell-like, co-operative

With written numbers:

thirty-five, eighty-seven, three-quarters, five thousand and twenty-nine

With a series of hyphenated words you can use your hyphens like this:

We used the three-, six-, and nine-second fuses

Always use hyphens when there is any danger of ambiguity:

re-cover, recover, re-creation, recreation, un-ionised, unionised.

When you are using compound adjectives you must use hyphens:

Little-used apparatus is a waste of space

means the apparatus is not used often.

Little used apparatus is a waste of space

means the apparatus is little and used. Take care that you hyphenate all the words you need to:

> The anti-animal experiments lobby are picketing the Body Shop factory

is rather different from:

> The anti-animal-experiments lobby are picketing the Biology Department

Hyphens can also be used to link two parts of a word split at the end of a line but it is best to avoid this as it can be confusing and looks bitty.

Apostrophes and inverted commas

The apostrophe

Apostrophes indicate possession, for example, *Vishal's ice bucket* and abbreviations, for example, *He's drunk.*

    With names ending in s, it is best to add *'s*, for example, *Mr Jones's coat*, but with classical names, for example, *Socrates*, it is often omitted, *Socrates' cat*. With plurals ending in s, for example, *thermometers,* you just write *thermometers'*, for example, *the thermometers' casings*. Where we have more than one noun we only use the apostrophe with the last noun, *Michael and Sue's wedding.*

    Apostrophes can also be used for abbreviations of plurals and numbers, *MSc's, MP's, I went to Tanzania in '98*. Apostrophes are also used in abbreviated words and phrases, for example, *don't doesn't, haven't*. Do not use these abbreviations in your dissertation or thesis. A problem, which crops up surprisingly often, is the confusion of *its* and *it's*. *Its* is the possessive. *It's* is an abbreviation of *it is*—do not, of course, abbreviate *it is* to *it's* in your text. It is very easy to slip apostrophes in by mistake and your spell-checker might not catch them, so run special checks for them using your 'FIND' command.

Inverted commas

The conventions about inverted commas, or quotation marks, vary from publisher to publisher, we shall follow the conventions of the Oxford University Press. Whatever set of conventions you choose to follow, make sure you are consistent. One point all publishers would agree on is that inverted commas are used to indicate quotes, written or spoken, to show what a person's actual words were. For example, if the actual words spoken were "I am on the verge of a breakthrough".

> The speaker said he 'was on the verge of a breakthrough'.

is wrong. Write:

> The speaker said he was 'on the verge of a breakthrough'.

or:

> The speaker said 'I am on the verge of a breakthrough'.

Inverted commas are also used to indicate colloquial or technical expressions, 'yo dude', 'goo-gum oscillator'. Inverted commas can also be used when quoting the title of a publication:

> Bowie, D. 1994. A logic based calculus of everyday objects. In 'Abducting Logic' O. Osborne (ed) (Hendrix and Joplin, Southampton) pp 3–10.

We do not use them for well-known books and publications such as the Bible, Koran, Talmud, Magna Carta, etc.

If you are using a short quote in your main text you can simply put it in inverted commas. It is best to indent longer quotes—do not use inverted commas with indented quotes.

In British English, if you have a quotation within a quotation use single inverted commas first, and then double inverted commas for the second quotation. If you have another quotation within your second use single inverted commas again, etc. In American English, the convention is reversed. In British English, if the quote comes at the end of a sentence, the full stop is put outside the inverted commas, in American English it is put inside. If a punctuation mark is part of the quotation keep it inside the inverted commas, and if the quotation ends in a full stop, question mark, or exclamation mark, you do not need to use another full stop after the inverted comma to end your sentence.

The slash

This is sometimes called a solidus /. It used to mean *per*, as in km/h. It should not be used to join two words together. Use a hyphen to do this. Do not write the solidus more than once in any notation.

The question mark

Keep these for direct questions:

> What is the secret of life?

With an indirect questions you do not use them:

> In this chapter we will ask what the secret of life is.

The exclamation mark

Over a pint of Bacardi in the pub, you might well say:

> Wow! I've discovered a new atom!

There really should not be any place in your dissertation or thesis for exclamation marks (other than as symbols in equations).

Accents

With words borrowed from other languages we keep whatever accents they came with, for example, *précis, pièce de résistance, papier-mâché, façade*.

Keep the accents on people's names when referencing them. It is worth playing around with your keyboard to find out how to get the different accents.

Although you will probably not be using it yourself, you might come across something called a dieresis, which is like the German *umlaut* ". It is used to clarify the pronunciation, for example, *naïve, coöperation,* and *zoölogy,* rather than *naive, co-operation* and *zoology.*

Common mistakes with punctuation

Fragments

Fragments are constructions like this:

> The blue mouse ate the cheese. Which had been left out by the kindly Mrs Karloff.

Which had been left out by the kindly Mrs Karloff is not a sentence in its own right. The sentence should read:

> The blue mouse ate the cheese which had been left out by the kindly Mrs Karloff.

or,

> The blue mouse ate the cheese, which had been left out by the kindly Mrs Karloff.

Comma splices

Comma-splices are constructions like this:

> The yellow mouse ate the blue cheese, it did not like the cheese much.

You cannot simply join two sentences with a comma. Either use a full stop:

> The yellow mouse ate the blue cheese. It did not like the cheese much.

or a conjunction, such as *and*, with or without a comma:

> The yellow mouse ate the blue cheese and it did not like the cheese much.

Run on sentences

A run on sentence is something like this:

> The yellow mouse ate the blue cheese it did not like the cheese much.

Here we have put what should be two sentences into one with no punctuation at all. Again we should use a full stop:

> The yellow mouse ate the blue cheese. It did not like the cheese much.

or a conjunction, with or without a comma.

> The yellow mouse ate the blue cheese and it did not like the cheese much.

Capitalisation

The conventions about capitalisation can be contradictory and illogical. Write *I* when writing about yourself, and start sentences and proper names (*John, Paris, Venezuela, Sir John Bodkin,* etc.) with a capital letter. Also capitalise the following:

Names of races, nationalities, languages, religions, and belief systems

> They are Asian. She is Thai. They are Colombian. He speaks Serbo-Croat. My family are Buddhist.

With regards to race and belief systems, some people hold that you should not capitalise *white* or *atheist*. If you happen to be writing about any kinds of deity in your thesis you can talk about a god, but God will normally be taken to mean the god of the Judeo-Christian religion; there is also the convention of referring to God as He, and occasionally as She.

Organisations, departments, jobs and areas of work or study

These are normally capitalised when they are used as proper names, the Army, the Labour Party, the Archbishop of York, Your Honour, His Excellency the Jordanian Ambassador, etc.:

> Bella Lugosi is Vice-Dean of Para-Psychology at the University of West Cheam. He is in a meeting the other vice-deans to discuss the recent crisis.
> Mr Ant is Head of Biochemistry at the University of West Cheam. He also teaches music and technical drawing.

so, for example, you would work in a chemistry department, but work in the Chemistry Department.

Names of periods of time and historical events

These should be capitalised if they are used as proper names, for example:

> Stone Age, Middle Ages, First World War.

Adjectives and nouns derived from names

These are capitalised if the connection to the name is still felt to be relevant:

Newtonian physics
Roman Empire
Aristotelian philosophy

However, we do not capitalise if the connection is felt to be remote:

french windows
arabic script
herculean effort
wellington boot
sandwich
jersey
boycott

or if it is an activity associated with a name:

galvanise
pasteurise

We do not capitalise scientific terms derived from names:

kelvin
joule
newton
henry

Although we do capitalise the unit abbreviation (see *Appendix 7: SI Units (Système International d'Unités) and Their Multiples*).

Titles of books and papers

There is a strong convention to capitalise the first letter of the words in titles, except for pronouns, prepositions and conjunctions unless they start the title:

The Adventures of Don Quixote

Remember that abbreviations which are not normally capitalised should never be capitalised in a title or heading; for example, the title 'RATE OF HZ INCREASE' is incorrect. It should read 'RATE OF Hz INCREASE', but titles are better not completely capitalised at all.

Sections of text

It is conventional to capitalise, for example, *Introduction* when you are using it to refer to your introduction. Do not capitalise it when referring to introductions in general. Similarly write Table 4, or Figure 5, but talk about figures and tables in general.

Beware of overusing capitals, particularly of CAPITALISING WHOLE WORDS OR SENTENCES. It is hard on the eyes and lacks subtlety. It is a bit like shouting at someone to try and get them to see your point of view—normally, you just irritate them. Look at any printed book and you will see that professional typesetters use capitals very sparingly.

Italics

Italics can be used as an alternative to inverted commas to highlight words in a text. Italics are also used to indicate words taken from other languages:

> *a priori*
> *in vitro*
> *in vivo*
> *in silico*

We do not normally italicise 'foreign' words that are in common usage, for example, et cetera, and you would not write 'My friends are on holiday in *Ibiza*.'

Italics can also be used as an alternative to inverted commas for titles in your references:

> Bowie, D. 1994. A logic based calculus of everyday objects. In *Abducting Logic*. Editor O. Osborne (Hendrix and Joplin, Southampton) pp 3–10.

Common Mistakes

Using the wrong tense.
Mixing tenses.
Misuse of words.
Vagueness.
Bad punctuation.

Key Points

Be clear, concise, and accurate.
Make sure what you write makes sense.
Make sure what you write means what you want it to.

Appendix 1

EASILY CONFUSED WORDS

to affect, to effect

To *affect* means to act upon something (also, to aim at, and to pretend).
To *effect* means to bring about, or accomplish. So, for example:

> Your strange behaviour *affects* your flatmates.

> You *effect* a breakthrough in chemistry.

an effect

An effect means something caused or produced.

to accept, to except

To accept is to receive willingly, *to except* is to exclude or make an exception.

advice, to advise

You *advise* someone (advise is the verb). They accept your *advice* (advice is the noun).

adviser, advisor

Both spellings are fine, although 'adviser' is more common in British English and 'advisor' more common in American English.

acknowledgement, acknowledgment

Acknowledgement is British English, *acknowledgment* is American English.

all ready, already

All ready means all of the things are ready. *Already* means by this time.

all right, alright

All right is considered 'more correct', so it probably best to use this in your thesis or dissertation.

anybody, anyone, anything

These are written as one word, unless you mean any body (just the body), one (thing), or thing.

to appreciate, to understand

Appreciate means to recognise the worth of or increase in value, *to understand* means comprehend.

around

Around means surrounding, do not use it instead of *about*.

around, round

These are interchangeable except in phrasal verbs like 'mess *around*', 'play *around*', and in phrases like 'all year *round*'.

artefact, artifact

Artefact is British English, *artifact* is American English.

as

As can mean either since, because or during.

assume, presume

These are more or less interchangeable, although *assume* tends to be used when one is not sure what one is saying is true, *presume* when one is stating what one believes to be true.

assure, ensure, insure

Assure - to give an assurance to remove doubt

Ensure - to make certain, to make sure

Insure - in British English this means to take out insurance; in American English it can also mean the same as ensure.

basal, basic

Basal is used mainly in technical and scientific writing, *basic* in more everyday English.

beside, besides

Beside means 'by the side of', or 'in comparison with'. *Besides* means 'in addition to', and 'other than'.

cannot

Cannot is usually written as one word.

can, may, could, might

Can expresses ability, 'I *can* speak French.' *May* expresses permission and possibility, 'You *may* leave.', 'This *may* be the room.' *Could* and *might* can be used either as the past tense of *can* and *may*, or in the present tense to express possibility, 'This *might/could* be the right room.'

compare with, compare to

When you want to indicate a similarity between things use *compare to*: 'Shall I *compare* thee *to* a summer's day?'

When you want to indicate the differences between things you can use either *to* or *with* 'Compared *with/to* London, Bristol is a small city.' *With* is always used when compare is in constructions such as 'Dogs *compare* favourably *with* cats when it comes to loyalty.'

complement, compliment

To have a *complement* means to have a complete set. A *compliment* is an expression of praise.

connection, connexion

Both are correct, although *connection* is the more usual spelling.

content, concentration

Content means amount. *Concentration* means weight per volume.

continual, continuous

Continual means frequently, *continuous* means without interruption.

could have, could of

'I *could have* come' is good English. 'I *could of* come' is nonsense and should never be used.

course, coarse

A *course* is what you study to gain a degree. *Coarse* means unrefined.

credible, creditable

Both mean believable, *creditable* also means worthy of praise.

criterion, criteria

Criterion is the singular, *criteria* is the plural.

data, datum

Datum is singular. *Data* is plural. So always write *data* in the plural, for example, 'the *data* are...', 'the *data* show...', 'the *data* were analysed...'.

different from, different than, different to

All are correct, although some people think that only *different from* should be used. *Different than* is more common in American than British English.

to defuse, to diffuse

To *defuse* is to remove a fuse, to *diffuse* is to disperse.

disc, disk

Disc tends to be more usual in British English, *disk* in American English. When referring to computer storage most people write *disk*, as we have done in this guide.

discreet, discrete

Discreet means cautious and circumspect. *Discrete* means distinct from.

disinterested, uninterested

Disinterested means impartial, with no interests in the issue. *Uninterested* means indifferent.

to disprove, to disapprove

To disprove mean to prove wrong, *disapprove* means to dislike.

Effective, effectual, efficacious, efficient

Effective means having the desired effect, coming into operation, or actually rather than theoretically existing.

Effectual means able to achieve the required effect—we only apply it to people in the negative: ineffectual.

Efficacious, is used for things (medicines, etc.) and means sure to produce the required effect.

Efficient means doing its job with the minimum of waste.

either/or, neither/nor

Either/or is used for positive comparisons, 'You can have *either* a banana *or* an apple.'

Neither/nor is used for negative comparison, 'He won *neither* fame *nor* fortune.'

elicit, illicit

Elicit means to get, for example, *to elicit* information from a supervisor.

Illicit means unlawful.

eligible, illegible

Eligible means fit to be chosen.

Illegible means unreadable.

to enquire, enquiry, to inquire, inquiry

In British English, *to inquire* and *inquiry* are used for formal investigations, *to enquire* and *enquiry* are used in the general sense. Americans tend to stick to *inquire* and *inquiry*.

farther, further

We use *farther* when talking about distance, *further* when talking about time or degree. 'Aberdeen is *farther* away from London than Brighton.', 'He waited for her a *further* two hours and thought *further* about the terrible secret he had uncovered, then he walked a little *farther*.'

few, fewer, fewest, little, less, least

We use *few* before countable nouns: *Few* people. *Fewer* people.

We use *little* before uncountable nouns: *Little* salt. *Less* salt.

By *a little* we mean a small amount. By *a few* we mean a small number.

foreword, forward

A *foreword* is an introduction to a book. *Forward* is a direction.

foreword, preface

The *foreword* is not written by the author of the book, the preface is.

got, gotten

In American English *gotten* is an equivalent of *got*. In British English *gotten* is not used.

gram, gramme

Gram is correct, do not use *gramme*. The abbreviation is simply g with no full stop.

if, whether

If is for possibilities, *whether* is for alternatives, 'I do not know *whether* (or not) my grant check has come.'

impracticable, impractical

Impracticable - cannot be carried out

Impractical - not practical, not sensible, not reasonable, etc.

into, in to

The student bumped *into* her supervisor, they walked *in to* discuss the project.

to lie, to lay

Lie, lying, lay, lain - mean to recline, 'The man *lies* down.'

Lay, laying, laid - mean (in non-colloquial English) to place or put something, 'John *laid* the pen on the desk.'

lighted, lit

Both are fine as the past tense of light.

longways, longwise, lengthways, lengthwise

Longways and *longwise* are both possible, but *lengthways* and *lengthwise* are more commonly used.

loose, lose

Loose is the opposite of tight. *Lose* is the opposite of win.

media, medium

Medium is singular, *media* is plural.

meantime, mean time, meanwhile, mean while

It is normal to write both as one word, *meantime, meanwhile*.

meter, metre

In British English, a *meter* is some kind of measuring instrument, and a *metre* is a measurement of length, in American English *meter* is sometimes used for both. However *metre* is the internationally agreed term for a measurement of length.

one, we

One can be used as an indefinite pronoun (also called impersonal and generic pronoun) to mean people in general, '*One* would not like to have to write *one's* thesis in Latin.' *We* can be used in a similar way, when the writer is including the reader, 'We need to keep an eye out for spelling mistakes.'

one of the best/worst etc.

This is fine if you mean there are a number of *best* or *worst* things (although one could argue that there can only be one *best* or *worst* thing). It is better to write: one of the better... one of the worse...

ought, ought to

Always use *to* with *ought*. 'Dr Karloff *ought to* be here by now.'

owing to, due to

Owing to means because of, '*Owing to* my total lack of experience I set fire to the High Energy Physics Laboratory.' *Due to* means as a result of. It needs a subject + verb in front of it, 'I set fire to the High Energy Physics Laboratory *due to* my total lack of experience.'

per cent, percent, percentage

Per cent is used in British English. American English tends to use *percent*. *Percentage* is a rate or proportion reckoned as so many *per cent*.

phenomena, phenomenon

Phenomena is plural. *Phenomenon* is singular.

to practise, practice

To practise is the verb. *A practice* is the noun.

to premise, premiss

To premise is to assume something from a *premiss*.

premises, premisses

Premises means a building. *Premisses* is the plural of *premiss*.

to prescribe, to proscribe

To prescribe is to lay down a rule. *To proscribe* is to ban.

principle, principal

A *principle* is a rule or accepted general truth. A *principal* is either a person in a position of authority or an adjective meaning the most important.

to proceed, to precede

To proceed means to go ahead with. *To precede* means to go before.

program, programme

Program is used in American English. In British English we use *programme*, except when writing about computers, when *program* tends to be used (as we have done in this guide).

proof, evidence

Proof is conclusive, *evidence* merely indicates that something might be true. Be very careful about claiming you have proved something.

proved, proven
These are both acceptable as past participles: 'It has been *proved*.', 'It has been *proven*.' *Proven* is more often used as an adjective, 'A *proven* idea.'

to quote, quotation
You *quote* a *quotation*.

to rise, to raise
Rise, rose, risen—mean to stand up or move upwards—either literally or metaphorically, 'Dr Karloff *rose* to the challenge.' *Raise, raised*, and *raised*, in British English, usually mean to make something move upwards, 'Dr Karloff *raised* his glass.' In American English *raise* can mean to make something grow, 'Dr Karloff has *raised* three landfish from birth'.

shall, will, should, would
Shall and *should* used to be used only with 'I', and 'we', but this distinction is no longer made.

sight, site, cite
Sight is the sense. *Site* is a place, as in building site. To *cite* is to quote or reference.

to sit, to set
Sit, sitting, sat, sat, mean to assume or maintain a sitting position, 'I *sat* on the chair.'
Set, setting, set, set, mean to put something in position, 'I *set* the cheese down on the plinth.'

stationary, stationery
Stationary means not moving. *Stationery* is paper, envelopes, etc.

systematic, systemic
Systematic is used to mean methodical. *Systemic* is used as a technical term meaning to do with how a system works.

times, fold, per cent
'*Times*', '*fold*', and '*per cent*' are often confused. We use them in the following ways: 'In the second experiment we used three *times* the concentration of NaCl.' but, 'In the second experiment the concentration was increased three*fold*.' If you write 'In the second experiment the concentration was increased three *times*.' you mean you increased the concentration on three different occasions, but give no information as to by how much you increased it. You could also write 'In the second experiment the concentration was increased by 300%.'

that, which, who, whom

You can use either *that*, *which*, *who*, or *whom* in a defining relative clause: 'The landfish *which/that* were put in the refrigerator have turned blue.', 'The woman *who/that* ate the mouse is in the cupboard.', 'The man *who/ whom/that* Dr Karloff shot has recovered.'

You can use *who*, *whom*, or *which* with non-defining relative clauses, 'The landfish, *which* were put in the refrigerator, have turned blue.', 'The woman, *who* ate the mouse, is in the cupboard.', 'The man, *who/whom* Dr Karloff shot, has recovered.'

We have covered defining and non-defining relative clauses in *Chapter 14: Use of English*.

to, too, two

To is the preposition, *too* means 'also' or indicates 'excess' as in '*too* much', *two* is the number.

toward, towards

Both are acceptable.

try to, try and

'To *try and* visit' means to try (to do something) and then to visit. 'To *try to* come' means just what it says. We have the same problem with *go and, come and* etc.

used to, use to

'I *used to* play rugby at school.' 'I *use* a pen *to* write my name.'

what ever, whatever, who ever, whoever, how ever, however, etc.

Write these as one word except in questions where the *ever* is used for emphasis: '*What ever* do you think you are doing?'

'*How ever* are we going to get out of this parking space?'

worth while, worth-while, worthwhile

In sentences like, 'The research is *worth while*.' write as two words. In sentences like 'It is *worth-while* research.' 'It is *worthwhile* research.' either hyphenate or write as one word.

whose, who's

Whose is the possessive of who, *who's* is an abbreviation of who is, which you should not be using in your thesis or dissertation.

Appendix 2

PREFIXES AND SUFFIXES

Sometimes prefixes will be joined to other words by hyphens, but usually this is only necessary to avoid ambiguity, for example, to distinguish between un-ionised and unionised—see the section on hyphens in *Chapter 14: The Use of English*.

Prefixes

ae, oe
Ligatured vowels like this are falling out of fashion. They are still printed squeezed together in Old English, French, Danish, Icelandic, and Norwegian words, but for general British English use they are printed as separate letters, for example, archaeology, larvae, paediatrics.

As time goes by the 'a' and 'o' are being dropped, so oecology and oeconomy and now spelt ecology and economy. British English uses ligatured vowels more than American English, so we have 'diarrhoea' (British), 'diarrhea' (American), ' oesophagus' (British), 'esophagus' (American).

anti-, ante-
The prefix *anti-* means against, *ante-* means before.

extra-
Extra means outside, for example, *extra-curricular* activities are those undertaken outside that which is taught as part of a curriculum or course.

inter-, intra-
The prefix *inter-* means between, among; *intra-* means within or on the inside.

Suffixes

-ae, -as
Most words ending in 'a' have a Latin root and a lot of them strictly take '-ae' as the plural, for example, 'larvae'. However, as our language evolves we tend to make a plural by putting an s onto the end of many of these words, for example, the plural of *nebula* is written as *nebulae* or *nebulas*. Some words now always end in an s if plural, for example, *areas*.

Words such as 'comma' and 'data' are already plural. If in doubt, consult a dictionary.

-able, -ible
Your spell-checker will pick out any mis-spellings you make of words that should take either *-able* or *-ible*, but there are some words that take different meanings in *-able* and
-ible, for example:

contractable	liable to be contracted as a disease or habit
contractible	capable of contracting or drawing together
forceable	able to be forced open
forcible	achieved by force
infusable	able to be infused
infusible	not able to be fused

other words can take either -able or -ible, for example:

collapsable, collapsible
extendable, extendible
extractable, extractible
preventable, preventible

Be guided by your dictionary.

Another quirk with -able is whether or not to omit the 'e' in some words like *'nameable'* or *'namable'*, *'saleable'* or *'salable'*, *'likeable'* or *'likable'*, *'useable'* or *'usable'*. Usually either is strictly possible, but often one spelling is more common than the other. Be guided by your dictionary and spell-checker.

-ative, -ive
Most words take the *-ative* suffix, for example, *'representative'* rather than *'representive'*. Be guided by your spell-checker and dictionary.

-ic, -ical
When making an adjective from a noun, some words always take *-ic*, *alcoholic*, some words always take *-ical*, *chemical*. Some words use either *-ic*, or *-ical*, but have different meanings, *economic, economical*. Some words differ between British and American English, for example, British English prefers *-ical geological*, American English tends to prefer *-ic geologic*.

-ion, -ment
Some words take *-ion* as a suffix, some take *-ment*. A few words take both but have different meanings, for example, *excitation* and *excitement*. Be guided by your dictionary.

-ion, -ness
Some words take *-ion* as a suffix, some take *-ness*. Some take both but have different meanings, *abstraction, abstractness, correction, correctness*. Be guided by your dictionary.

-ist, -alist
Some words can take either, for example, *educationist, educationalist, horticulturist, horticulturalist*. Be guided by your dictionary and spell-checker.

-ise, -ize
Some words always use *-ise, advertise, advise, hybridise*. Other words take either -ise or -ize. American English always goes for the -ize option, for example, *hybridize*. Be consistent with which one you use; if you are submitting a thesis or dissertation to a British university or college, it is probably wisest to use *-ise*.

-os, -oes
There are no rules as to whether all words ending in *o* should have an *-oes* ending in the plural or an *-os* plural, so we get *potatoes,* and *Eskimos;* be guided by your spell-checker and dictionary.

-or, -er
Some words end in *-or*, some words end *-er*, for example, teacher and distributor. On the whole, Latin-based words tend to end in -or; be guided by your spell-checker and dictionary.

Appendix 3

WORDY WORDS AND PHRASES

Try to remove these words and phrases from your text. They are usually unnecessary.

above
Above can cause wordiness, for example,

> The argument experiment outlined *above* proved crucial to our research.

You could just write,

> This proved crucial to our research.

Below is a little more useful. You can use it as a signpost for related information or argument the reader will find useful, for example, *I will deal with this contradiction below.*

amount
'A large *amount*' is vague, and 'a maximum' *amount* is redundant—maximum and minimum are amounts.

all, all of
Try to get rid of the *'of'*, *'all the samples'* is more concise than *'all of the samples'*.

area

My research was in the *area* of delta wing aeronautical design.

could be written more concisely,

My research was in delta wing aeronautical design.

as far as

As far as bore holes are concerned, I had no problems.

You could just write,

I had no problems with bore holes.

both, both of

Both of the traces indicated an increase in activity.

This does not need the *of* and could be written,

Both traces indicated an increase in activity.

capability

The plastic has the *capability* of reforming itself.

You could just write,

The plastic can reform itself.

cause and result

At times you have to make clear what is a *cause* and what is a *result*, but there are often quicker ways of saying what you mean; compare the following, for example,

The addition of NaCl *caused* an improvement in taste.
Adding NaCl improved the taste.

clearly demonstrates, shows

If data *clearly demonstrate* a phenomenon, then they really *show* it.

> The spectrophotometer readings *clearly demonstrate* a decrease in density.
> The spectrophotometer readings *show* a decrease in density.

definitely

Unless there is some question hanging over what you are saying, *definitely* is unnecessary and often looks desperate to be convincing.

due to the fact that

This really means 'because'...

in addition

If you are adding something to your text, you do not need to tell the reader you are doing so by using *'in addition'*.

in colour, in appearance

Both of these phrases are unnecessary,

> The landfish was red in colour.
> The landfish was red in appearance.

could simply be written as,

> The landfish was red.

in order to

Use 'to'.

literally

A word best avoided in your thesis. You whole text should be literal.

manner

Manner always makes your sentence wordy.

> The chemical were added in a slow *manner*.
> The chemicals were added slowly. [This example is also vague.]

nature
> Chewing gum has an elastic *nature*.
> Chewing gum is elastic.

process
Unless you are actually discussing a *process* this is a word to avoid. For example, you do not have to write 'the stratification *process*', you can just write 'stratification'.

personally
Personally I think...

This is not only wordy but can deflate your argument. It implies that other people would disagree with you. *I think* does the job better.

pooled together
If you pool samples, then they must be together, so you could write,

> The sulphide samples were pooled together.

as simply,

> The sulphide samples were pooled.

reason, because
You only need one of them, as they do the same job.

seldom ever
You do not need the *ever*.

similar, very similar
The *very* is redundant. Writing that two items are *very similar* tells us no more than if we write they are *similar*.

sized
Large *sized*...
If something is large, we know this refers to size, so the word *sized* is redundant. [This example is also vague.]

that

One *that* in a sentence is normally more than enough unless you want to stress a point; for example, the first and second sentences are clear, the third sentence is full of redundant *that*s:

> I found I could not move, wanted to go to the toilet, had a headache, and had fallen down the stairs.

> I found *that* I could not move, wanted to go to the toilet, had a headache, and had fallen down the stairs.

> I found *that* I could not move, *that* I wanted to go to the toilet, *that* I had a headache, and *that* I had fallen down the stairs.

the field of

This is useful when you are giving the general area in which you work,

> My research was in *the field of* Para-psychology.

but it can get overused,

> *The field of* para-psychology is seldom taken seriously by other scientists.

could equally well be written,

> Para-psychology is seldom taken seriously by other scientists.

Appendix 4

WORDS THAT CAN CAUSE VAGUENESS

This list is not exhaustive, but it contains the most common words you should look out for in your text and avoid using if possible.

Quality and manner

brilliant
excellent
fantastic
good
lovely
marvellous
wonderful

awful
bad
dreadful
terrible
worse
worst

beautifully
best
better
brilliantly
excellently

fantastically
marvellously
well
wonderfully

awfully
badly
dreadfully
terribly

cold
cool
hot
warm

Size

big
huge
large
little
long
massive
short
small
tiny

Amount

a few
few
least
less
little
many
most
more
much
some

Speed

fast
quick
quickly
speedily
swift
swiftly

slow
slowly

Time

at once
lately
recently
soon

Frequency

frequently
occasionally
often
periodically
sometimes
usually

hardly ever
rarely
scarcely
seldom

Degree

almost
barely
enough

extremely
fairly
far
hardly
little
much
nearly
quite [unless you mean completely]
rather
really
scarcely
so
too
very

great
high

little
low
negligible
poor

greatly
highly

lowly
poorly

Appendix 5

LATIN WORDS AND ABBREVIATIONS, AND WHAT THEY MEAN

We have given the meanings of some words, phrases, and abbreviations that you may come across in scientific as well as everyday writing. We have not always given the literal translation; sometimes we have given the current day meaning. Plain English is best for most writing, including scientific writing, so do not use these words and phrases unless they are appropriate. Non-English words are usually italicised, apart from those in common use, for example, generally *in vivo* would be italicised, but etc. would not.

ab extra   from the outside
ab initio   from the beginning
'The project has been under-funded *ab initio*, and therefore has gone slowly.'
ab intra   from within, from the inside
addendum   a thing to be added
Addenda is the plural of *addendum*
ad hoc   to this, for this specific purpose
This refers to something temporary, something just used for a specific and time-limited purpose. 'An *ad hoc* election was held and Majid was voted in as the graduate student in charge of fetching the beer at the meeting.'
ad infinitum   forever
'Margaret felt as if her chemistry tutorial stretched on *ad infinitum* into the future; she resolved to study law instead.'
am, a.m.   morning
Ante meridiem, this actually means before noon.

a priori   from what is known
'We have no *a priori* knowledge of how many genes are involved in Down syndrome.'

appendix   something added
Appendices is the plural of appendix

bona fide   in good faith, honestly, the real thing
'He is a *bona fide* representative from the chemicals company, and not an insurance salesman.'

c., ca., circa   about
Use *circa* when you are not sure about a date. 'We believe that the Head of Department was born *c.* 1930.'

cf.   compare
cf. is an abbreviation for *confer*, which is Latin for compare. 'I think Dr Lloyd's guitar riffs are the best in the Department (but *cf.* Dr Beck's drum solo).'

corrigenda   items for correction
You may come across a *corrigenda* when a paper is being corrected at the proof stage. It is simply of a list of things to correct.

de facto   in reality
'Dr Egan is the *de facto* Head of Department, because his boss, Professor Smith, spends so much time bird-watching.'

de novo   from new, afresh
'We were not happy with the quality of the phosphoglycerol supply and so we decided to synthesise it again, *de novo*, from new stocks of chemicals.'

diem perdidi   another day wasted
Actually, you are unlikely to ever see this written down in anything scientific, but we all know the feeling.

e.g.   for example
This is an abbreviation of *exempli gratia*, which literally means 'for the sake of example'. The abbreviation, e.g., is not normally italicised.

ergo   therefore
These days it is not normal to use this in a scientific text.

errata   list of errors
You may encounter such a list when correcting a scientific paper. *Errata* is simply the plural of *erratum*, meaning error or mistake.

et al.   and others
Et al. is an abbreviation for either *et alii*, or *et aliae*, or *et alia*, which mean and other men, women, and things, respectively. Use *et al.* to mean 'and

others' and so avoid long lists, such as lists of authors. 'The paper by Souper *et al.* is the result of a collaboration between thirty different laboratories.'

etc.   the rest, and so on

This is an abbreviation for two words, *et cetera*, which mean 'and the next'.

ibid., ibidem   in the same place

You are unlikely to come across this in scientific writing, but we have included it just in case. It means refer to the identical source as the previous one, in other words, look at the same reference as has just been cited.

id., idem   the same

Again, this is unlikely to crop up in scientific literature, but just so that you know, when an author is cited many times, the abbreviation *id.* is used in place of the author's name, after the first citation of that name.

i.e., *id est*   that is

'Their white lab coats were stained with loading dye, i.e., the stains were blue.'

infra   below, underneath

Compare infra red with ultra violet.

in silico   by computer

The genes were found *in silico* by analysing the DNA database with the latest software.

in situ   in the natural location

In situ is generally used to indicate that the material you are working with was studied in its natural location. '*In situ* hybridisation techniques allow us to visualise the chromosomes within a cell nucleus'.

inter   among, or between

Use inter to refer to something between individual entities. 'There is a lot of inter-University rivalry between the Oxford and Cambridge boat race teams.' 'The intermolecular distances are very high because the molecules repel each other.' Do not confuse inter and intra. Inter is normally used as a prefix, and is not italicised.

in toto   totally, completely

'The cost of the analyses, including all extras, comes to £28,000 *in toto*.'

intra   within

Use 'intra' to refer to something within an individual entity. 'Intramolecular forces keep the atoms together within the crystal.' 'It is sometimes hard to find polymorphisms from two strains in an intra-species cross.' 'Christmas can be a chore because of intra-family rows.' Do not confuse inter and intra. Intra is normally used as a prefix, and is not italicised.

in utero   in the uterus

This phrase is refers to the embryo or fetus in the uterus. '*In utero* surgery on the fetus is hard to perform but can correct some disorders.'

in vacuo   in a vacuum

This phrase is most often used literally by physicists, chemists and engineers, and can also be used metaphorically by the rest of us to mean 'in isolation'—avoid this usage in your text.

in vitro   in the test tube (or your equivalent)

This phrase is literally translated as 'in glass'. We use it for when referring to studies taking place in glass (or plastic) containers.

in vivo   in the living organism

'Sally and Ian are studying the depolarisation of the heart, *in vivo*.'

locus   place

'The gene maps to this locus on the chromosome (Clifford and Sergot, 1999).'

N.B.   take note

N.B. is an abbreviation of *nota bene*, which means 'note well'. This abbreviation should not find its way into your writing because it is too informal, and suggests that you have not properly explained something.

non sequitur   it does not follow

A non sequitur is a conclusion that does not logically follow from the given premiss.

p.a., per annum   each year, yearly

'The bench fees are £13,000 p.a.'

per capita   per head, individually

'The charge, per capita, for the lab's day trip to Brighton is £25.'

per diem   daily, each day

'The charge, per diem, for Professor Jackson's stay in the hotel is £180'.

per se   by, or in itself

'I am not discussing Professors *per se*, I am talking about academics in general.'

premiss   set before

A statement from which something is inferred.

pm, p.m.   afternoon

Post meridiem, actually means after noon.

post mortem   after death, autopsy

'Dr Martin performed a post mortem on the body.' Post mortem is a phrase in common usage and does not need to be italicised.

post partem   after parting, immediately after childbirth

'Isolation in a high rise flat may exacerbate post partem depression.'

QED   thus it is demonstrated

An abbreviation for the Latin *quod erat demonstrandum*. Do not use this in your scientific writing, it is archaic and very few people will know what you mean.

q.v.   which see

Quod vide, *q.v.*, is a way of referring you elsewhere in the text. 'A cow is a large quadruped (*q.v.*)'. This sentence is telling you that you can look up the word 'quadruped' in the text if you need to.

sic   so, thus

Sic is used to draw attention to a mistake in a quotation. This can either be a spelling mistake, a factual mistake or a mistake in usage. *Sic* is usually placed inside square brackets. 'The newspaper article stated that domestic salt was soddum [*sic*] chloride and vinegar was ascetic [*sic*] acid.' 'She remarked that the night sky was beautiful and planet was the most voluminous [*sic*] she'd seen.' 'He wrote that crossing the dessert [*sic*] by camel was arduous.'

status quo   the state things are in

Status quo refers to the current conditions. 'By paying scientists higher salaries we may alter the *status quo* of society.'

ultra   beyond

Compare ultra violet with infra red.

vice versa   the other way round, the position being reversed, conversely

'Dr Simpson was an excellent collaborator for Dr Fisher, and vice versa.'

viva voce   oral examination

Viva voce literally translates as 'with the living voice', but has come to refer to an oral exam.

viz.   namely

Viz is an abbreviation of *videlicet,* which literally translates as 'it is permitted to see', and is used to mean 'namely'.

Appendix 6

THE GREEK ALPHABET AND ROMAN NUMERALS

The Greek Alphabet

capital	small letters	name
A	α	alpha
B	β	beta
Γ	γ	gamma
Δ	δ	delta
E	ε	epsilon
Z	ζ	zeta
H	η	eta
Θ	θ	theta
I	ι	iota
K	κ	kappa
Λ	λ	lambda
M	μ	mu
N	ν	nu
Ξ	ξ	xi
O	ο	omicron
Π	π	pi
P	ρ	rho
Σ	σ	sigma
T	τ	tau
Y	υ	upsilon
Φ	φ	phi

K	κ	chi
Ψ	ψ	psi
Ω	ω	omega

Roman Numerals

1	I
2	II
3	III
4	IV
5	V
6	VI
7	VII
8	VIII
9	IX
10	X
11	XI
12	XII
13	XIII
14	XIV
15	XV
16	XVI
17	XVII
18	XVIII
19	XIX
20	XX
21	XXI
30	XXX
40	XL
50	L
60	LX
70	LXX
80	LXXX
90	XC
100	C
101	CI

110	CX
199	CIC
200	CC
400	CD
500	D
900	CM
1000	M
1500	MD
1600	MDC
1700	MDCC
1900	MCM
1997	MCMXCVII
1998	MCMXCVIII
1999	MCMXCIX
2000	MM

Appendix 7

SI UNITS (SYSTÈME INTERNATIONAL D'UNITÉS) AND THEIR MULTIPLES

The SI units are the internationally standardised units of measurement. The system has seven base units (m, kg, s, A, K, cd, mol) and two supplementary units (rad, sr). All the other units are derived from these.

Each unit has a symbol. The symbols are not capitalised unless the measurement is named after a person, in which case the first letter of the symbol is capitalised. Do not use full stops with the abbreviations of SI units. Multiples of the units are given in decimals (for example, micro 10^{-6}, milli 10^{-3}, kilo 10^3, mega 10^6). The only commonly used decimals are those that are a multiple of three.

	Name	Symbol
length	metre	m
mass	kilogram	kg
time	second	s
electric current	ampere	A
thermodynamic temperature	kelvin	K
luminous intensity	candela	cd
amount of substance	mole	mol
plane angle	radian	rad
solid angle	steradian	sr
frequency	hertz	Hz
energy	joule	J
force	newton	N

power	watt	W
pressure	pascal	Pa
electric charge	coulomb	C
electric potential difference	volt	V
electric resistance	ohm	W
electric conductance	siemens	S
electric capacitance	farad	F
magnetic flux	weber	Wb
inductance	henry	H
magnetic flux density (magnetic induction)	tesla	T
luminous flux	lumen	lm
illuminance, (illumination)	lux	lx
absorbed dose	gray	Gy
activity	becquerel	Bq
dose equivalent	sievert	Sv

Multiples to be used with SI Units

multiple	prefix	symbol
10^{-18}	atto-	a
10^{-15}	femto-	f
10^{-12}	pico-	p
10^{-9}	nano-	n
10^{-6}	micro-	m
10^{-3}	milli-	m
10^{-2}	centi-	c
10^{-2}	deci-	d
10^{1}	deca-	da
10^{2}	hecto-	h
10^{3}	kilo-	k
10^{6}	mega-	M
10^{9}	giga-	G
10^{12}	tera-	T
10^{15}	peta-	P
10^{18}	exa-	E

INDEX